Advance Praise for *The End of Growth*

Heinberg draws in the big three drivers of inevitable crisis—resource constraints, environmental impacts, and financial system overload—and explains why they are not individual challenges but one integrated systemic problem. By time you finish this book, you will have come to two conclusions. First, we are not facing a recession—this is the end of economic growth. Second, this is not our children's problem—it is ours. It's time to get ready, and reading this book is the place to start.

— PAUL GILDING, author, *The Great Disruption*,
Former head of Greenpeace International

Richard has rung the bell on the limits to growth. This is real. The consequences for economics, finance, and our way of life in the decades ahead will be greater than the consequences of the industrial revolution were for our recent ancestors. Our coming shift from quantity of consumption to quality of life is the great challenge of our generation—frightening at times, but ultimately freeing.

— JOHN FULLERTON, President and Founder, Capital Institute

Why have mainstream economists ignored environmental limits for so long? If Heinberg is right, they will have a lot of explaining to do. The end of conventional economic growth would be a shattering turn of events—but the book makes a persuasive case that this is indeed what we are seeing.

— LESTER BROWN, Founder, Earth Policy Institute
and author, *World on the Edge*

Heinberg shows how peak oil, peak water, peak food, etc. lead not only to the end of growth, and also to the beginning of a new era of progress without growth.

— HERMAN E. DALY, Professor Emeritus,
School of Public Policy, University of Maryland

The End of Growth offers a comprehensive, timely and persuasive analysis of the reality of ecological limits as they relate to economic growth. Filled with facts and figures and very readable, the book makes a rational case while paying attention to nuance and counterarguments. A must-read for anyone who depends upon economic growth, which means all of us.

— LESLIE E. CHRISTIAN, CFA, President and CEO Portfolio 21 Investments

Heinberg has masterfully summarized and updated the case against economics, and its fraudulent scorecard—GDP. He explains why conventional economic growth is ending now, and why growth of human populations and material consumption will follow suit. Yet we all can still grow in wisdom and continue expanding the knowledge of our universe, while growing greener technologies capturing the sun's daily free photon flow as we transition to the Solar Age.

— HAZEL HENDERSON, author, *The Politics of the Solar Age* (1981) and other books, President of Ethical Markets Media (USA and Brazil) and its Green Transition Scoreboard®

Dig into this book! It is crammed full of ideas, information and perspective on where our troubled world is headed—a Baedeker for the perplexed, and that's most of us.

— JAMES GUSTAVE SPETH, author of *The Bridge at the Edge of the World: Capitalism, the Environment and Crossing from Crisis to Sustainability*

Read this book and have the light switched on.

— CAROLINE LUCAS, Member of Parliament (UK)

Richard Heinberg is not one to shy away from difficult topics and The *End of Growth* is no exception. Heinberg explains today's environmental and economic realities—which are scary to face. But believe me, not facing them is a whole lot scarier. And as Heinberg explains, the sooner we have this critically needed conversation about how to live in a healthy, fair, and meaningful way on this one planet we have, the better it will be for all of us.

—ANNIE LEONARD, author, *The Story of Stuff*

A vitally important book—it helps clear away many of the mistaken assumptions that clutter our heads when we think about 'obvious' and 'natural' facts of our economic life. You really need to read it if you want to understand the next few crucial years.

— BILL MCKIBBEN, author of *Deep Economy* and *Eaarth*

From all my research, I'm come to appreciate how much the expectation of unending growth dominates public policy — and how ephemeral that goal is likely to prove. Until now, however, no one has had the foresight to address this critical topic. Congratulations to Richard Heinberg for providing such a lucid account of the natural limits to growth and the urgent need for a new economic model.

— MICHAEL KLARE, author, *Rising Powers, Shrinking Planet*

THE END
of
GROWTH

post carbon **institute**

The End of Growth is a "living book."
Additional materials are being posted and updated
regularly at endofgrowth.com.

Also, you can join the conversation about EOG at
facebook.com/richardheinberg
facebook.com/postcarbon

DEDICATION

*To P. J., whose generosity makes my work possible —
and who keeps going even with the knowledge
that time and means work against us.*

THE END
of
GROWTH

Adapting to Our New Economic Reality

RICHARD
HEINBERG

NEW SOCIETY PUBLISHERS

Cover design by Diane McIntosh.
Balloon: iStock © kutay tanir. Pin: iStock © Keith Webber Jr.
Background texture: iStock © toto8888

Printed in Canada. Second printing October 2011.

Paperback ISBN: 978-0-86571-695-7
eISBN: 978-1-55092-483-1

Inquiries regarding requests to reprint all or part of *The End of Growth*
should be addressed to New Society Publishers at the address below.

To order directly from the publishers, please call toll-free
(North America) 1-800-567-6772, or order online at www.newsociety.com

Any other inquiries can be directed by mail to:

New Society Publishers
P.O. Box 189, Gabriola Island, BC V0R 1X0, Canada
(250) 247-9737

New Society Publishers' mission is to publish books that contribute in fundamental
ways to building an ecologically sustainable and just society, and to do so with the
least possible impact on the environment, in a manner that models this vision.
We are committed to doing this not just through education, but through action.
Our printed, bound books are printed on Forest Stewardship Council-certified
acid-free paper that is **100% post-consumer recycled** (100% old growth forest-
free), processed chlorine free, and printed with vegetable-based, low-VOC inks,
with covers produced using FSC-certified stock. New Society also works to reduce
its carbon footprint, and purchases carbon offsets based on an annual audit to
ensure a carbon neutral footprint. For further information, or to browse our full
list of books and purchase securely, visit our website at: www.newsociety.com

Library and Archives Canada Cataloguing in Publication

Heinberg, Richard
    The end of growth : adapting to our new economic reality / Richard Heinberg.

Includes index.
ISBN 978-0-86571-695-7

1. Economic indicators. 2. Economic forecasting.
3. Economic development. 4. Natural resources. I. Title.

HC59.3.H45 2011   330.9'05   C2011-902953-7

NEW SOCIETY PUBLISHERS
www.newsociety.com

Contents

Acknowledgments

This book has benefited from the contributions of many people whose help deserves recognition.

I will start with the staff of Post Carbon Institute, the organization for which I am happy and proud to work. Asher Miller, Daniel Lerch, Ken White, and Tod Brilliant read the manuscript in various stages and offered key editorial suggestions. This team has also collaborated with me on related projects—including short YouTube videos such as the award-winning "300 Years of Fossil Fuels in 300 Seconds" and "Who Killed Growth?" Crystal Santorineos transcribed interviews and Simone Osborne published excerpts from the book on PostCarbon.org and Energy Bulletin.net (readers of those excerpts in turn sent in valuable suggestions for the manuscript).

New Society Publishers also deserve thanks—especially Ingrid Witvoet, who edited the manuscript; Chris and Judith Plant, who responded enthusiastically to *The End of Growth* when it was no more than an idea; Sue Custance who helped shepherd the project along; and E. J. Hurst, who masterminded promotional efforts.

Jared Finnegan put his recent studies at the London School of Economics to use in offering informed editorial suggestions for the manuscript, and also in preparing the graphs and their captions—a task that involved gathering dozens of data sets from a variety of sources. Jared's editorial input was especially crucial to the section "The End of 'Development'" in Chapter 5.

Suzanne Doyle contributed many hours to compiling the Notes, finding sources and keeping track of hundreds of references.

My research for *The End of Growth* entailed interviews and conversations with many brilliant thinkers, each of whom deserves an introduction as well as my great thanks.

A conversation with Gus Speth—one of the giants of the environmental movement, who taught for many years at Yale (and now at Vermont Law School), and author of *A Bridge at the Edge of the World: Capitalism, the Environment, and Crossing from Crisis to Sustainability*—brought me up to speed with recent developments in the world of alternative economics.

Another conversation, this one with John Fullerton, formerly Managing Director of JPMorgan and more recently the founder of the Capital Institute, helped me understand the culture of Wall Street and how it evolved during the past quarter century. These days John is working on steering the investment world toward a just, sustainable, and resilient economy and is pioneering several promising efforts in that direction.

Herman Daly, pioneer ecological economist, former World Bank economist, and author of *The Growth Illusion*, generously agreed to read and comment on Chapter 6.

Nate Hagens, hedge fund manager-turned-ecological economist, read most of the manuscript and contributed many important suggestions; his expert knowledge of the workings of financial markets informed Chapter 2.

Josh Farley, Fellow of the Gund Institute for Ecological Economics and a Professor in the Community Development and Applied Economics faculty at the University of Vermont (and co-author with Herman Daly of the textbook *Ecological Economics*), read most of the manuscript and provided crucial guidance in several areas. A few of the explanatory Notes are lifted nearly verbatim from his comments in the margins of the manuscript.

Hazel Henderson, futurist and author of *Ethical Markets: Growing the Green Economy*, has been critiquing conventional economic theory for decades while offering alternative ways of making money work for people (rather than the other way around). Interviewing her helped open my thinking to possibilities I hadn't considered, and I tried to capture these in Chapters 6 and 7.

Chris Martenson, creator of "The Crash Course" and veteran of ten years in corporate finance and strategic consulting, writes an ongoing series of commentaries about the world financial situation. Because

Chris's worldview is shaped by his awareness of resource limits and by systems thinking, I find his analysis particularly credible and useful. His insights are reflected in Chapter 2 and 3.

Nicole Foss is co-editor of TheAutomaticEarth.com, where she writes under the name Stoneleigh; she also runs the Agri-Energy Producers' Association of Ontario, Canada, where she focuses on farm-based bio-gas projects and grid connections for renewable energy. While living in the UK, Nicole was a Research Fellow at the Oxford Institute for Energy Studies, where she specialized in nuclear safety in Eastern Europe and the former Soviet Union. I've benefited from reading many of her commentaries on the world financial scene, and Chapter 2 incorporates a number of important insights she conveyed during a long conversation in December 2010.

Charles Hall, Professor of Systems Ecology at State University of New York College of Environmental Science and Forestry in Syracuse, is the world's principal authority on the subject of net energy, discussed in Chapter 4. While I owe a general, long-standing debt to Charlie's research (whose importance is still unrecognized by many energy analysts), his most recent peer-reviewed publications provide crucial support to some of the core arguments in this book.

David Murphy, one of Charlie Hall's former students and now an important researcher in his own right, is the author and co-author of key peer-reviewed papers on net energy cited in Chapter 4. David read that chapter and made important suggestions.

Michael Klare, Professor of Peace Studies at Hampshire College and author of *Rising Powers, Shrinking Planet*, is the world's foremost authority on linkages between resource depletion and geopolitics. A conversation with him in late 2010 was the basis for the section "Post-Growth Geopolitics" in Chapter 5.

Bill Ryerson, founder and President of Population Media Center and President of the Population Institute, is the wisest, best informed, and most effective advocate for population issues I know. Our conversation laid the groundwork for the section "Population Stress: Old vs. Young on a Full Planet" in Chapter 5.

Mats Larsson, a management consultant and author of several books including *The Limits of Business Development and Economic Growth*, was an important source for Chapter 4, which he also read and commented on.

Jason Bradford, a former research biologist at Washington University and Missouri Botanical Garden, who currently leads the farmland management program for Farmland LP, read and commented upon much of the manuscript and drafted a sidebar for Chapter 7.

Warren Karlenzig, one of the world's top experts on urban sustainability strategy and metrics, and author of *How Green is Your City? The SustainLane U.S. City Rankings*, is a frequent traveler to China, where he consults with local authorities on planning issues. His insight into that nation's predicament and prospects is reflected in the section "The China Bubble" in Chapter 5.

David Fridley, staff scientist at the Energy Analysis Program at Lawrence Berkeley National Laboratory and expert on China's energy policy, also offered important input to "The China Bubble."

Doug Tompkins, who enjoyed notable success in business and went on to become one of the world's leading conservationists, read the manuscript and made helpful suggestions, while the Foundation for Deep Ecology provided material assistance for the promotion of this book.

Helena Norberg-Hodge, founder of the International Society for Ecology and Culture, has been a friend and inspiration for many years; her perspective is explicit in the section "The End of 'Development'" in Chapter 5, but also subtly flavors Chapters 1 and 6.

These expert readers and sources kept me from making many mistakes that would otherwise have compromised the core message of *The End of Growth*. Any errors that remain are my sole responsibility.

I am also deeply indebted to the work of Dennis Meadows and Jorgen Randers — the surviving members of the original Limits to Growth research group. If the world had listened, today we would all have much less to worry about.

Finally, I would like once again to thank my wife Janet Barocco for her tireless support and encouragement, and for helping make our home a place of artistry, good humor, and natural beauty.

Introduction: The New Normal

Leading active members of today's economics profession…have formed themselves into a kind of Politburo for correct economic thinking. As a general rule — as one might generally expect from a gentleman's club — this has placed them on the wrong side of every important policy issue, and not just recently but for decades. They predict disaster where none occurs. They deny the possibility of events that then happen…. They oppose the most basic, decent and sensible reforms, while offering placebos instead. They are always surprised when something untoward (like a recession) actually occurs. And when finally they sense that some position cannot be sustained, they do not reexamine their ideas. They do not consider the possibility of a flaw in logic or theory. Rather, they simply change the subject. No one loses face, in this club, for having been wrong. No one is dis-invited from presenting papers at later annual meetings. And still less is anyone from the outside invited in.

— James K. Galbraith (economist)

The central assertion of this book is both simple and startling: *Economic growth as we have known it is over and done with.*

The "growth" we are talking about consists of the expansion of the overall size of the economy (with more people being served and more money changing hands) and of the quantities of energy and material goods flowing through it.

The economic crisis that began in 2007–2008 was both foreseeable and inevitable, and it marks a *permanent, fundamental* break from past

1

decades — a period during which most economists adopted the unrealistic view that perpetual economic growth is necessary and also possible to achieve. There are now fundamental barriers to ongoing economic expansion, and the world is colliding with those barriers.

This is not to say the US or the world as a whole will never see another quarter or year of growth relative to the previous quarter or year. However, when the bumps are averaged out, the general trend-line of the economy (measured in terms of production and consumption of real goods) will be level or downward rather than upward from now on.

Nor will it be impossible for any region, nation, or business to continue growing for a while. Some will. In the final analysis, however, this growth will have been achieved at the expense of other regions, nations, or businesses. From now on, only *relative growth* is possible: the global economy is playing a zero-sum game, with an ever-shrinking pot to be divided among the winners.

Why Is Growth Ending?

Many financial pundits have cited serious troubles in the US economy — including overwhelming, un-repayable levels of public and private debt, and the bursting of the real estate bubble — as immediate threats to economic growth. The assumption generally is that eventually, once these problems are dealt with, growth can and will resume at "normal" rates. But the pundits generally miss factors *external* to the financial system that make a resumption of conventional economic growth a near-impossibility. *This is not a temporary condition; it is essentially permanent.*

Altogether, as we will see in the following chapters, there are three primary factors that stand firmly in the way of further economic growth:

- The *depletion* of important resources including fossil fuels and minerals;
- The proliferation of *negative environmental impacts* arising from both the extraction and use of resources (including the burning of fossil fuels) — leading to snowballing costs from both these impacts themselves and from efforts to avert them; and
- *Financial disruptions* due to the inability of our existing monetary, banking, and investment systems to adjust to both resource scarcity and soaring environmental costs — and their inability (in the context

of a shrinking economy) to service the enormous piles of government and private debt that have been generated over the past couple of decades.

Despite the tendency of financial commentators to ignore environmental limits to growth, it is possible to point to literally thousands of events in recent years that illustrate how all three of the above factors are interacting, and are hitting home with ever more force.

Consider just one: the Deepwater Horizon oil catastrophe of 2010 in the US Gulf of Mexico.

The fact that BP was drilling for oil in deep water in the Gulf of Mexico illustrates a global trend: while the world is not in danger of *running out* of oil anytime soon, there is very little new oil to be found in onshore areas where drilling is cheap. Those areas have already been explored and their rich pools of hydrocarbons are being depleted. According to the International Energy Agency, by 2020 almost 40 percent of world oil production will come from offshore. So even though it's hard, dangerous, and expensive to operate a drilling rig in a mile or two of ocean water, that's what the oil industry must do if it is to continue supplying its product. That means more expensive oil.

Obviously, the environmental costs of the Deepwater Horizon blowout and spill were ruinous. Neither the US nor the oil industry can afford another accident of that magnitude. So, in 2010 the Obama administration instituted a deepwater drilling moratorium in the Gulf of Mexico while preparing new drilling regulations. Other nations began revising their own deepwater oil exploration guidelines. These will no doubt make future blowout disasters less likely, but they add to the cost of doing business and therefore to the already high cost of oil.

The Deepwater Horizon incident also illustrates to some degree the knock-on effects of depletion and environmental damage upon financial institutions. Insurance companies have been forced to raise premiums on deepwater drilling operations, and impacts to regional fisheries have hit the Gulf Coast economy hard. While economic costs to the Gulf region were partly made up for by payments from BP, those payments forced the company to reorganize and resulted in lower stock values and returns to investors. BP's financial woes in turn impacted British pension funds that were invested in the company.

This is just one event — admittedly a spectacular one. If it were an isolated problem, the economy could recover and move on. But we are, and will be, seeing a cavalcade of environmental and economic disasters, not obviously related to one another, that will stymie economic growth in more and more ways. These will include but are not limited to:

- Climate change leading to regional droughts, floods, and even famines;
- Shortages of energy, water, and minerals; and
- Waves of bank failures, company bankruptcies, and house foreclosures.

Each will be typically treated as a special case, a problem to be solved so that we can get "back to normal." But in the final analysis, they are all related, in that they are consequences of a growing human population striving for higher per-capita consumption of limited resources (including non-renewable, climate-altering fossil fuels), all on a finite and fragile planet.

Meanwhile, the unwinding of decades of buildup in debt has created the conditions for a once-in-a-century financial crash — which is unfolding around us, and which on its own has the potential to generate substantial political unrest and human misery.

The result: we are seeing a perfect storm of converging crises that together represent a watershed moment in the history of our species. We are witnesses to, and participants in, the transition from decades of economic growth to decades of economic contraction.

The End of Growth Should Come As No Surprise

The idea that growth will stall out at some point this century is hardly new. In 1972, a book titled *Limits to Growth* made headlines and went on to become the best-selling environmental book of all time.[1]

That book, which reported on the first attempts to use computers to model the likely interactions between trends in resources, consumption, and population, was also the first major scientific study to question the

State of the World

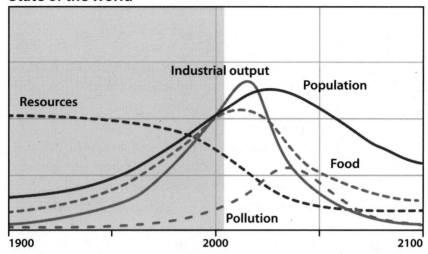

FIGURE 1. Limits to Growth Scenario. Source: *The Limits to Growth: The 30-Year Update* (2004), p. 169.

assumption that economic growth can and will continue more or less uninterrupted into the foreseeable future.

The idea was heretical at the time — and still is. The notion that growth *cannot* and *will not* continue beyond a certain point proved profoundly upsetting in some quarters, and soon *Limits to Growth* was prominently "debunked" by pro-growth business interests. In reality, this "debunking" merely amounted to taking a few numbers in the book completely out of context, citing them as "predictions" (which they explicitly were not), and then claiming that these predictions had failed.[2] The ruse was quickly exposed, but rebuttals often don't gain nearly as much publicity as accusations, and so today millions of people mistakenly believe that the book was long ago discredited. In fact, the original *Limits to Growth* scenarios have held up quite well. (A recent study by Australian Commonwealth Scientific and Industrial Research Organization (CSIRO) concluded, "[Our] analysis shows that 30 years of historical data compares favorably with key features of [the *Limits to Growth*] business-as-usual scenario....")[3]

The authors fed in data for world population growth, consumption trends, and the abundance of various important resources, ran their computer program, and concluded that the end of growth would probably arrive between 2010 and 2050. Industrial output and food production would then fall, leading to a decline in population.

The *Limits to Growth* scenario study has been re-run repeatedly in the years since the original publication, using more sophisticated software and updated input data. The results have been similar each time.[4]

Why Is Growth So Important?

During the last couple of centuries, economic growth became virtually the sole index of national well-being. When an economy grew, jobs appeared and investments yielded high returns. When the economy stopped growing temporarily, as it did during the Great Depression, financial blood-letting ensued.

Throughout this period, world population increased — from fewer than two billion humans on planet Earth in 1900 to over seven billion today; we are adding about 70 million new "consumers" each year. That makes further economic growth even more crucial: if the economy stagnates, there will be fewer goods and services *per capita* to go around.

We have relied on economic growth for the "development" of the world's poorest economies; without growth, we must seriously entertain the possibility that hundreds of millions — perhaps billions — of people will never achieve the consumer lifestyle enjoyed by people in the world's industrialized nations. From now on, efforts to improve quality of life in these nations will have to focus much more on factors such as cultural expression, political freedoms, and civil rights, and much less on an increase in GDP.

Moreover, we have created monetary and financial systems that *require* growth. As long as the economy is growing, that means more money and credit are available, expectations are high, people buy more goods, businesses take out more loans, and interest on existing loans can be repaid.[5] But if the economy is not growing, new money *isn't* entering the system, and the interest on existing loans cannot be paid; as a result, defaults snowball, jobs are lost, incomes fall, and consumer spending contracts — which leads businesses to take out fewer loans, causing still less new money to enter the economy. This is a self-reinforcing destructive feedback loop that is very difficult to stop once it gets going.

In other words, the existing market economy has no "stable" or "neutral" setting: there is only growth or contraction. And "contraction" can be just a nicer name for recession or depression — a long period of cascading job losses, foreclosures, defaults, and bankruptcies.

We have become so accustomed to growth that it's hard to remember that it is actually is a fairly recent phenomenon.

Over the past few millennia, as empires rose and fell, local economies advanced and retreated — while world economic activity overall expanded only slowly, and with periodic reversals. However, with the fossil fuel revolution of the past century and a half, we have seen economic growth at a speed and scale unprecedented in all of human history.[6] We harnessed the energies of coal, oil, and natural gas to build and operate cars, trucks, highways, airports, airplanes, and electric grids — all the essential features of modern industrial society. Through the one-time-only process of extracting and burning hundreds of millions of years' worth of chemically stored sunlight, we built what appeared (for a brief, shining moment) to be a perpetual-growth machine. We learned to take what was in fact an extraordinary situation for granted. It became *normal*.

But as the era of cheap, abundant fossil fuels comes to an end, our assumptions about continued expansion are being shaken to their core. The end of growth is a very big deal indeed. It means the end of an era, and of our current ways of organizing economies, politics, and daily life.

It is essential that we *recognize and understand the significance of this historic moment*: if we have in fact reached the end of the era of fossil-fueled economic expansion, then efforts by policy makers to continue pursuing elusive growth really amount to a flight from reality. World leaders, if they are deluded about our actual situation, are likely to delay putting in place the support services that can make life in a nongrowing economy tolerable, and they will almost certainly fail to make needed, fundamental changes to monetary, financial, food, and transport systems.

As a result, what could be a painful but endurable process of adaptation could instead become history's greatest tragedy. We can survive the end of growth, and perhaps thrive beyond it, but only if we recognize it for what it is and act accordingly.

BOX I.1   But Isn't the US Economy Recovering?

From July 2009 through the end of 2010, the US economy posted GDP gains — i.e., signs of growth. Nominal GDP surpassed pre-recession levels in mid-2010, while inflation-adjusted GDP nearly returned to its pre-recession level.[7] This followed GDP contraction in the months December 2007 through June 2009.[8]

But, as we will see in Chapter 6, GDP is a poor gauge of overall economic health. Even if GDP has returned to former levels, the economy of the United States is fundamentally changed: unemployment levels are much higher and tax revenues for state and local governments are severely reduced. Some economists may define this technically as a recovering and growing economy, but it certainly is not a healthy one.

Moreover, much of this apparent growth has come about because of enormous injections of stimulus and bailout money from the Federal government. Subtract those, and the GDP growth of the past year or so almost disappears.

On the basis of historical analysis of previous financial crises, economists Carmen Reinhart and Kenneth Rogoff conclude that the economic crisis of 2008 will have

"...deep and lasting effects on asset prices, output and employment. Unemployment rises and housing price declines extend out for five and six years, respectively. On the encouraging side, output declines last only two years on average. Even recessions sparked by financial crises do eventually end, albeit almost invariably accompanied by massive increases in government debt.... The global nature of the [current] crisis will make it far more difficult for many countries to grow their way out through higher exports, or to smooth the consumption effects through foreign borrowing. In such circumstances, the recent lull in sovereign defaults is likely to come to an end."[9]

But this analysis considers only the financial aspects of the crisis and ignores the deeper issues of energy, resources, and environment. The "recovery" that began in 2009 occurred in the context of energy prices

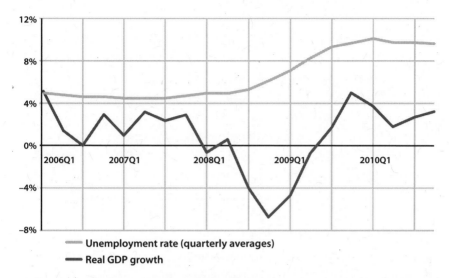

FIGURE 2. Economic Growth and Unemployment, 2006–2010. As the US economy contracted from the financial crisis in 2008, economic growth went negative and the unemployment rate shot up. Source: US Bureau of Labor Statistics, US Bureau of Economic Analysis.

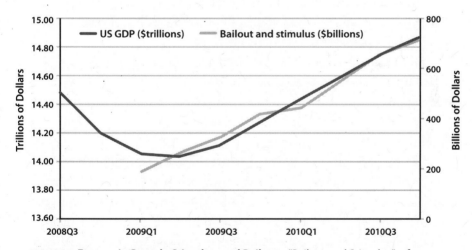

FIGURE 3. Economic Growth, Stimulus, and Bailouts. "Bailout and Stimulus" refers to the Troubled Asset Relief Program (TARP) and the American Recovery and Reinvestment Act of 2009. As this graph shows, these federal government expenditures appear to have been the primary source of economic growth since the financial crisis in 2008. What happens when the federal government can no longer bail out the banks and stimulate the economy? Source: US Bureau of Economic Analysis, The Committee for a Responsible Federal Budget.

that had fallen substantially from their peak in mid-2008; but as consumer demand showed tepid signs of revival in late 2010, oil prices lofted upward again. If this "recovery" continues, energy prices will rise even further and contraction will resume.

In short: while the US economy may have posted growth (as technically defined) in 2009–2010, it is operating in a fundamentally different mode than before: it is led to a greater extent than before by government spending (as opposed to consumer activity), and it is hostage to energy prices.

But Isn't Growth Normal?

Economies are systems, and as such they follow rules analogous (to a certain extent) to those that govern biological systems. Plants and animals tend to grow quickly when they are young, but then they reach a more or less stable mature size. In organisms, growth rates are largely controlled by genes, but also by availability of food.

In economies, growth seems tied to the availability of resources, chiefly energy ("food" for the industrial system), and credit ("oxygen" for the economy) — as well as to economic planning.

During the past 150 years, expanding access to cheap and abundant fossil fuels enabled rapid economic expansion at an average rate of about three percent per year; economic planners began to take this situation for granted. Financial systems internalized the expectation of growth as a promise of returns on investments.

Most organisms cease growing once they reach adulthood; if curtailment of growth weren't genetically programmed, plants and animals would outgrow a range of practical constraints: imagine, for example, the survival challenges faced by a two-pound hummingbird. If the analogy holds, then economies must eventually stop growing too. Even if planners (society's equivalent of regulatory DNA) dictate more growth, at some point increasing amounts of "food" and "oxygen" will cease to be available. It is also possible for wastes to accumulate to the point that the biological systems that underpin economic activity (such as forests, crops, and human bodies) are smothered and poisoned.

But many economists don't see things this way. That's probably because current economic theories were formulated during the anomalous historical period of sustained growth that is now ending. Economists are merely generalizing from their experience: they can point to decades of steady growth in the recent past, and they simply project that experience into the future.[10] Moreover, they have theories to explain why modern market economies are immune to the kinds of limits that constrain natural systems: the two main ones have to do with *substitution* and *efficiency*.

If a useful resource becomes scarce, its price will rise, and this creates an incentive for users of the resource to find a substitute. For example, if oil gets expensive enough, energy companies might start making liquid fuels from coal. Or they might develop other energy sources undreamed of today. Many economists theorize that this process of substitution can go on forever. It's part of the magic of the free market.

Boosting efficiency means doing more with less. In the US, the number of dollars generated in the economy for every unit of energy consumed has increased steadily over recent decades.[11] Part of this increasing efficiency is a result of outsourcing manufacturing to other nations — which must then burn the coal, oil, or natural gas to make our goods. (If we were making our own running shoes and LCD TVs, we'd be burning that fuel domestically.)[12] Economists also point to another, related form of efficiency that has less to do with energy (in a direct way, at least): the process of identifying the cheapest sources of materials, and the places where workers will be most productive or work for the lowest wages. As we increase efficiency, we use less — of energy, resources, labor, or money — to do more. That enables more economic growth.

Finding substitute resources and upping efficiency are undeniably effective adaptive strategies of market economies. Nevertheless, the question remains as to how long these strategies can continue to work in the real world — which is governed less by economic theories than by the laws of physics. In the real world, some things don't have substitutes, or the substitutes are too expensive, or don't work as well, or can't be produced fast enough. And efficiency follows a law of diminishing returns: the first gains in efficiency are usually cheap, but every further incremental gain tends to cost more, until further gains become prohibitively expensive.

In the end, we can't outsource more than 100 percent of manufacturing, we can't transport goods with zero energy, and we can't enlist the efforts of workers and count on their buying our products while paying them nothing. Unlike most economists, most physical scientists recognize that growth within any functioning, bounded system has to stop sometime.

BOX 1.2   **Cooking the Books on Growth**

Are government economic statistics accurate and credible? Not according to consulting economist John Williams of shadowstats.com. After a "lengthy process of exploring the history and nature of economic reporting and interviewing key people involved in the process from the early days of government reporting through the present," Williams began compiling his own data and publishing them on his website. In some cases, as with unemployment statistics, he simply highlights the discrepancy between current definitions and reporting practices and former ones: if unemployment numbers were reported today the way they were in the 1970s, the current figure would be in the range of 16–18 percent rather than the officially reported 9–10 percent (for example, people who have given up looking for jobs are no longer categorized as "unemployed").

"Shadow stats" for inflation are consistently higher than the government's reported figures, and GDP growth rates consistently lower.

Regarding Figure 4, Williams notes, "The SGS-Alternate GDP reflects the inflation-adjusted, or real, year-to-year GDP change, adjusted for distortions in government inflation usage and methodological changes that have resulted in a built-in upside bias to official reporting."

All of which raises the question: How much of the economic "recovery" is actually only smoke and mirrors?

The Simple Math of Compounded Growth

In principle, the argument for an eventual end to growth is a slam-dunk. If any quantity grows steadily by a certain fixed percentage per year, this implies that it will double in size every so-many years; the higher the percentage growth rate, the quicker the doubling. A rough method of figuring

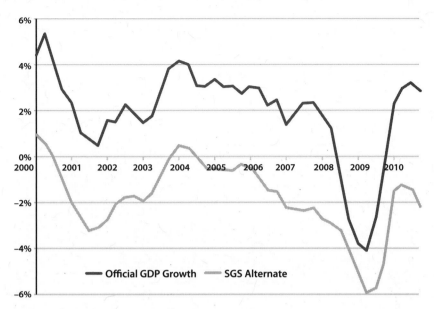

FIGURE 4. US GDP Growth, Official vs. Shadowstats, 2000–2010. Official GDP data comes from the Bureau of Economic Analysis. The SGS Alternate comes from Shadow Government Statistics. Both datasets are adjusted for inflation. Source: Shadow Government Statistics, American Business Analytics and Research LLC, shadowstats.com

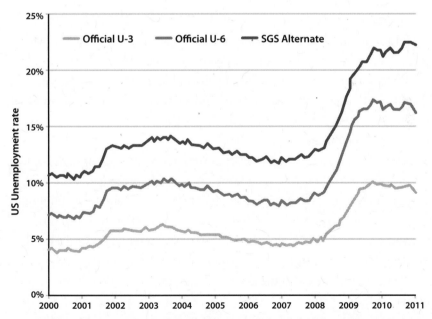

FIGURE 5. Civilian Unemployment, Official vs. Shadowstats, 2000–2010 (Seasonally Adjusted). The SGS-Alternate Unemployment Rate reflects current unemployment reporting methodology adjusted for the significant portion of "discouraged workers" no longer included after 1994. The Bureau of Labor Statistics U-6 rate includes both discouraged workers as currently defined (discouraged less than one year) and long-term discouraged workers (discouraged more than one year). Source: Shadow Government Statistics, American Business Analytics and Research LLC, shadowstats.com.

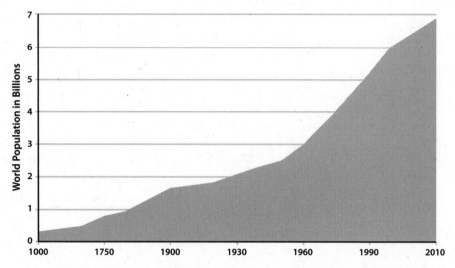

FIGURE 6. World Population Growth, 1000–2010. Source: Population Division of the Department of Economic and Social Affairs of the United Nations Secretariat, "World Population Prospects: The 2008 Revision" (2009–10 population data based on 2008 projection).

doubling times is known as the rule of 70: dividing the percentage growth rate into 70 gives the approximate time required for the initial quantity to double. If a quantity is growing at 1 percent per year, it will double in 70 years; at 2 percent per year growth, it will double in 35 years; at 5 percent growth, it will double in only 14 years, and so on. If you want to be more precise, you can use the Y^x button on a scientific calculator, but the rule of 70 works fine for most purposes.

Here's a real-world example: Over the past two centuries, human population has grown at rates ranging from less than one percent to more than two percent per year. In 1800, world population stood at about one billion; by 1930 it had doubled to two billion. Only 30 years later (in 1960) it had doubled again to four billion; currently we are on track to achieve a third doubling, to eight billion humans, around 2025. No one seriously expects human population to continue growing for centuries into the future. But imagine if it did — at just 1.3 percent per year (its growth rate in the year 2000). By the year 2780 there would be 148 trillion humans on Earth — one person for each square meter of land on the planet's surface.

It won't happen, of course.

In nature, growth always slams up against non-negotiable constraints sooner or later. If a species finds that its food source has expanded, its numbers will increase to take advantage of those surplus calories—but then its food source will become depleted as more mouths consume it, and its predators will likewise become more numerous (more tasty meals for them!). Population "blooms" (or periods of rapid growth) are nearly always followed by crashes and die-offs.[13]

Here's another real-world example. In recent years China's economy has been growing at eight percent or more per year; that means it is more than doubling in size every ten years. Indeed, China now consumes more than twice as much coal as it did a decade ago—the same with iron ore and oil. The nation now has four times as many highways as it did, and almost five times as many cars. How many more doublings can occur before China has used up its key resources—or has simply decided that enough is enough and has stopped growing? The question is hard to answer with a specific number, but it is unlikely to be a large one.

This discussion has very real implications, because the economy is not just an abstract concept; it is what determines whether we live in luxury or poverty, whether we eat or starve. If economic growth ends, everyone will be impacted, and it will take society years to adapt to this new condition. Therefore it is important to know whether that moment is close at hand or distant in time.

The Peak Oil Scenario

As mentioned, this book will argue that global economic growth is over because of a convergence of three factors—resource depletion, environmental impacts, and systemic financial and monetary failures. However, a single factor may be playing a key role in bringing the age of expansion to a close. That factor is oil.

Petroleum has a pivotal place in the modern world—in transportation, agriculture, and the chemicals and materials industries. The Industrial Revolution was really the Fossil Fuel Revolution, and the entire phenomenon of continuous economic growth—including the development of the financial institutions that facilitate growth, such as fractional reserve banking—is ultimately based on ever-increasing supplies of cheap energy.

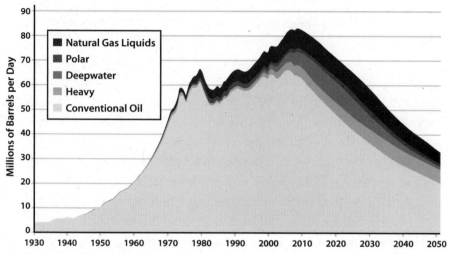

FIGURE 7. World Oil Production. Source: Colin Campbell, personal comunication.

Growth requires more manufacturing, more trade, and more transport, and those all in turn require more energy. This means that if energy supplies can't expand and energy therefore becomes significantly more expensive, economic growth will falter and financial systems built on expectations of perpetual growth will fail.

As early as 2000, petroleum geologist Colin Campbell discussed a Peak Oil impact scenario that went like this.[14] Sometime around the year 2010, he theorized, stagnant or falling oil supplies would lead to soaring and more volatile oil prices, which would precipitate a global economic crash. This rapid economic contraction would in turn lead to sharply curtailed energy demand, so oil prices would then fall; but as soon as the economy regained strength, demand for petroleum would recover, prices would again soar, and as a result of that the economy would relapse. This cycle would continue, with each recovery phase being shorter and weaker, and each crash deeper and harder, until the economy was in ruins. Financial systems based on the assumption of continued growth would implode, causing more social havoc than the oil price spikes would themselves directly generate.

Meanwhile, volatile oil prices would frustrate investments in energy alternatives: one year, oil would be so expensive that almost any other energy source would look cheap by comparison; the next year, the price of oil would have fallen far enough that energy users would be flocking back to it, with investments in other energy sources looking foolish. But low oil prices would discourage exploration for more petroleum, leading to even worse fuel shortages later on. Investment capital would be in short supply in any case because the banks would be insolvent due to the crash, and governments would be broke due to declining tax revenues. Meanwhile, international competition for dwindling oil supplies might lead to wars between petroleum importing nations, between importers and exporters, and between rival factions within exporting nations.

In the years following the turn of the millennium, many pundits claimed that new technologies for crude oil extraction would increase the amount of oil that can be obtained from each well drilled, and that enormous reserves of alternative hydrocarbon resources (principally tar sands and oil shale) would be developed to seamlessly replace conventional oil, thus delaying the inevitable peak for decades. There were also those who said that Peak Oil wouldn't be much of a problem even if it happened soon, because the market would find other energy sources or transport options as quickly as needed — whether electric cars, hydrogen, or liquid fuel made from coal.

In succeeding years, events appeared to be supporting the Peak Oil thesis and undercutting the views of the oil optimists. Oil prices trended steeply upward — and for entirely foreseeable reasons: discoveries of new oilfields were continuing to dwindle, with most new fields being much more difficult and expensive to develop than ones found in previous years. More oil-producing countries were seeing their extraction rates peaking and beginning to decline despite efforts to maintain production growth using high-tech, expensive extraction methods like injecting water, nitrogen, or carbon dioxide to force more oil out of the ground. Production decline rates in the world's old, super-giant oilfields, which are responsible for the lion's share of the global petroleum supply, were accelerating. Production of liquid fuels from tar sands was expanding only slowly, while the development of oil shale remained a hollow promise for the distant future.[15]

From Scary Theory to Scarier Reality

Then in 2008, the Peak Oil scenario became all too real. Global oil production had been stagnant since 2005 and petroleum prices had been soaring upward. In July 2008, the per-barrel price shot up to nearly $150 — half again higher (in inflation-adjusted terms) than the price spikes of the 1970s that had triggered the worst recession since World War II. By summer 2008, the auto industry, the trucking industry, international shipping, agriculture, and the airlines were all reeling.

But what happened next riveted the world's attention to such a degree that the oil price spike was all but forgotten: in September 2008, the global financial system nearly collapsed. The most frequently discussed reasons for this sudden, gripping crisis had to do with housing bubbles, lack of proper regulation of the banking industry, and the over-use of bizarre financial products that almost nobody understood. However, the oil price spike had also played a critical (if largely overlooked) role in initiating the economic meltdown.[16]

In the immediate aftermath of that global financial near-death experience, both the Peak Oil impact scenario proposed a decade earlier and

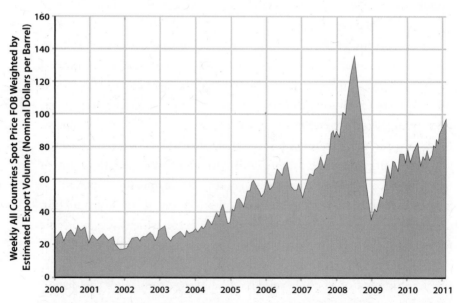

FIGURE 8. World Crude Oil Prices, 2000–2011. Source: US Energy Information Administration.

the *Limits to Growth* standard-run scenario of 1972 seemed to be confirmed with uncanny and frightening accuracy. Global trade was falling. The world's largest auto companies were on life support. The US airline industry had shrunk by almost a quarter. Food riots were erupting in poor nations around the world. Lingering wars in Iraq (the nation with the world's second-largest crude oil reserves) and Afghanistan (the site of disputed oil and gas pipeline projects) continued to bleed the coffers of the world's foremost oil-importing nation.[17]

Meanwhile, the dragging debate about what to do to rein in global climate change exemplified the political inertia that had kept the world on track for calamity since the early '70s. It had by now become obvious to a great majority of people familiar with the scientific data that the world has two urgent, incontrovertible reasons to rapidly end its reliance on fossil fuels: the twin threats of climate catastrophe and impending constraints to fuel supplies. Yet at the landmark international Copenhagen climate conference in December 2009, the priorities of the most fuel-dependent nations were clear: carbon emissions should be cut, and fossil fuel dependency reduced, *but only if doing so does not threaten economic growth.*

Bursting Bubbles

As we will see in Chapters 1 and 2, expectations of continuing growth had in previous decades been translated into enormous amounts of consumer and government debt. An ever shrinking portion of America's wealth was being generated by invention of new technologies and manufacture of consumer goods, and an ever greater portion was coming from buying and selling houses, or moving money around from one investment to another.

As a new century dawned, the world economy lurched from one bubble to the next: the emerging-Asian-economies bubble, the dot-com bubble, the real estate bubble. Smart investors knew that these would eventually burst, as bubbles always do, but the smartest ones aimed to get in early and get out quickly enough to profit big and avoid the ensuing mayhem.

If Peak Oil and other limits on resources were closing the spigots on growth in 2007–2008, the pain that ordinary citizens were experiencing seemed to be coming from other directions entirely: loss of jobs and collapsing real estate prices.

In the manic days of 2002 to 2006, millions of Americans came to rely on soaring real estate values as a source of income, turning their houses into ATMs (to use once more the phrase heard so often then). As long as prices kept going up, homeowners felt justified in borrowing to remodel a kitchen or bathroom, and banks felt fine making those loans. Meanwhile, the wizards of Wall Street were finding ways of slicing and dicing sub-prime mortgages into tasty collateralized debt obligations that could be sold at a premium to investors — with little or no risk! After all, real estate values were destined to just keep going up. *God's not making any more land*, went the truism.

Credit and debt expanded in the euphoria of easy money. All this giddy optimism led to a growth of jobs in construction and real estate industries, masking underlying ongoing job losses in manufacturing.

A few dour financial pundits used terms like "house of cards," "tinder-box," and "stick of dynamite" to describe the situation. All that was needed was a metaphoric breeze or rogue spark to produce a catastrophic outcome. Arguably, the oil price spike of mid-2008 was more than enough to do the trick.

But the housing bubble was itself merely a larger fuse: in reality, the entire economic system had come to depend on impossible-to-realize expectations of perpetual growth and was set to detonate. Money was tied to credit, and credit was tied to assumptions about growth. Once growth went sour in 2008, the chain reaction of defaults and bankruptcy began; we were in a slow-motion explosion.

Since then, governments have worked hard to get growth started again. But, to the very limited degree that this effort temporarily succeeded in late 2009 and 2010, it did so by ignoring the underlying contradiction at the heart of our entire economic system — the assumption that we can have unending growth in a finite world.

What Comes After Growth?

The realization that we have reached the point where growth cannot continue is undeniably depressing. But once we have passed that psychological hurdle, there is some moderately good news. The end of economic growth does not necessarily mean we've reached the end of qualitative improvements in human life.

Not all economists have fallen for the notion that growth will go on forever. There are schools of economic thought that recognize nature's limits; and, while these schools have been largely ignored in policy circles, they have developed potentially useful plans that could help society adapt.

The basic factors that will inevitably shape whatever replaces the growth economy are knowable. To survive and thrive for long, societies have to operate within the planet's budget of sustainably extractable resources. This means that even if we don't know in detail what a desirable post-growth economy and lifestyle will look like, we know enough to begin working toward them.

We must discover how life in a non-growing economy can actually be fulfilling, interesting, and secure. The absence of growth does not necessarily imply a lack of change or improvement. Within a non-growing or equilibrium economy there can still be continuous development of practical skills, artistic expression, and certain kinds of technology. In fact, some historians and social scientists argue that life in an equilibrium economy can be superior to life in a fast-growing economy: while growth creates opportunities for some, it also typically intensifies competition — there are big winners and big losers, and (as in most boom towns) the quality of relations within the community can suffer as a result. Within a non-growing economy it is possible to maximize benefits and reduce factors leading to decay, but doing so will require pursuing appropriate goals: instead of *more*, we must strive for *better*; rather than promoting increased economic activity for its own sake, we must emphasize that which increases quality of life without stoking consumption. One way to do this is to reinvent and redefine *growth* itself.

The transition to a no-growth economy (or one in which growth is defined in a fundamentally different way) is inevitable, but it will go much better if we plan for it rather than simply watch in dismay as institutions we have come to rely upon fail, and then try to improvise a survival strategy in their absence.

In effect, we have to create a desirable "new normal" that fits the constraints imposed by depleting natural resources. *Maintaining the "old normal" is not an option*; if we do not find new goals for ourselves and plan our transition from a growth-based economy to a healthy equilibrium economy, we will end up with a much less desirable "new normal." Indeed,

we are already beginning to see this in the forms of persistent high unemployment, a widening gap between rich and poor, and ever more frequent
and worsening environmental crises—all of which translate to profound
distress across society.

A Guide to the Book

This book began with a sudden insight on the morning of September 16,
2008 (the day after Lehman Brothers filed for bankruptcy). I was sitting in
a meeting of about 40 leaders and funders of non-profit organizations, lis-
tening to a former JP Morgan managing director explain what derivatives
are and why the financial world seemed to be disintegrating at that very
moment. One of the funders in the room took a call on his cell phone and
afterward I heard him whisper, "I just lost forty million dollars." The notion occurred to me: *We are witnessing the beginning of the end of economic
growth.* I knew the end was inevitable anyway, but now events within the
world of high finance were conspiring with environmental limits to bring
it about sooner, and more dramatically, than almost anyone had foreseen.

That thought wouldn't have stayed with me if I hadn't been prepared
for it—conditioned by having read *the Limits to Growth* decades previously, and by years of following trends in resource depletion. But it did
take root, and for months afterward I poked and prodded it every which
way, testing to see if it was sound, premature, or plain wrong.

I discussed it with economists, business consultants, energy experts,
and resource analysts. I spent countless hours reading about economic
history and about the causes of the unfolding financial catastrophe. I
consulted my colleagues at Post Carbon Institute, asking: Even if this is
true—that the world has indeed essentially outgrown the possibility of
growth itself—is this a message that should be broadcast to the world, or
would it be better for me to continue writing about energy and resource
issues? At last, in mid-2010, for reasons I'll discuss more in Chapter 7, it
became clear that the story of *The End of Growth* needed to be told.

The realization that growth may be at an end raises many questions.
Will the financial impact be inflationary or deflationary? Will some nations fare better than others, leading to protectionist trade wars? Will the
"downsizing" of the economy lead also to a "downsizing" of the human

species? How quickly will all of this happen? What can we do to protect ourselves and adapt?

These are some of the issues we will explore in the chapters ahead.

Chapter 1 is a potted history of economies and the discipline of economics. Readers well-versed in these subjects will find this a quick and dirty tour. This is not because I lack formal training as an economist or historian (though I do), but because the purpose here is only to provide some context. The rest of the book assumes a basic understanding of how and why economies have come to rely on growth, and why most mainstream economic theories ignore environmental limits.

In Chapter 2 we will see why economic growth has stumbled badly for reasons internal to the world's monetary and financial systems. Crucially, we will explore whether there are practical limits to debt, and whether we have broached those limits. This chapter also provides a short history of the current worldwide economic crisis and the efforts of governments and central banks to manage the mayhem.

Chapter 3 examines factors external to the financial system that will make it impossible for the economy to recover and begin growing again — factors that include the depletion of fossil fuels, minerals, and other natural resources, as well as worsening natural and industrial disasters.

Many readers will protest that limits to energy resources and minerals can be overcome with efficiency and substitution, enabling further economic growth. Chapter 4 addresses those arguments, showing why economic strategies that worked well to maintain an expansive trajectory during the 20th century are losing steam.

Chapter 5 explores how the winding down of world economic growth is likely to play out over the coming decades in terms of demography, international development, currency wars, and geopolitical rivalries. This chapter also addresses China's continued rapid economic expansion and examines in some detail the question: Can this continue for long?

In Chapter 6 we will explore ways that governments and central banks could successfully manage the inevitable transition from a growth-dependent economy to a contracting or steady-state economy. We begin the chapter with a rather stark portrayal of a "default scenario" of what is likely to transpire if the managers of the global money system continue

with current policies. Along the way, we learn about alternative currencies, ecological economics, and the economics of happiness.

Finally, Chapter 7 discusses what individuals and communities can do
now to prepare for changed conditions ahead, laying the groundwork for
the post-growth, post-hydrocarbon economy and way of life. As hopeful
signs and opportunities, we explore Transition Initiatives and Common
Security Clubs.

I recommend reading these chapters in sequence. The book develops
its argument cumulatively.

The process of writing *the End of Growth* changed me. Even though
I was well prepared to undertake the project, having spent the past four
decades observing how and why our current growth-based economy is
unsustainable, I found the process of coming to terms with the implications of an ongoing cessation of worldwide economic expansion more
than sobering. Even readers well versed in relevant subjects such as ecological economics will likely find that this book undermines their mental
equilibrium in a way that is both deeply uncomfortable and exhilarating—
in that it makes explicit a host of fears and misgivings about the economy
that I think most of us carry around with us unconsciously.

| BOX 1.3 | **The Perils of Prediction** |

This book is in effect making a prediction—that world economic growth
will not return. It is a hedged prediction, because it takes account of
the likelihood of relative growth, consisting of temporarily continuing
expansion in some economies and occasional partial rebounds in others. Still, hedged or not, predictions are perilous in fields ranging from
weather forecasting to horse racing, economics certainly among them.[18]

Some would argue that timing is the essence of prediction.[19] If a
forecast is off by a few years (or even milliseconds, in some scientific
experiments), the prediction fails. Paul Ehrlich was famously wrong in
his 1980 bet with Julian Simon that the prices of five commodity metals
would increase over the following decade. Arguably, Ehrlich just had his
timing wrong: as we have seen, since 2000 most commodity prices have
trended upward. But by calling the commodity price rise for too soon he

lost the $10,000 bet and provided resource optimists with an endlessly repeatable anecdote.

Others would say that, at least in predictive situations that involve a dire warning, the general correctness of the warning is often more important than the precise timing specified. Suppose the National Hurricane Center forecasts that a hurricane will strike Miami at approximately 5 pm, but the storm's speed across water slows temporarily and the hurricane actually strikes at 11 pm, still wreaking devastation. The important thing will have been that people were warned and got out of harm's way; the forecasters' failure to pinpoint the moment of impact will be seen to have been of little importance — it did not make the hurricane disappear.

The end of growth is a process, and, as I hope to have successfully argued, it is an inevitable one. The crash of 2008 was undoubtedly a pivotal moment in that process, but the shift from a general pattern of economic growth to one of general contraction is likely to continue for several years. Relative growth will make confirmation or disconfirmation of the prediction implied in this book's title problematic during this time. However, the real aim of the book is not to score points for accuracy in forecasting an event that must occur in any case (whether it happens this year or a decade from now), but to warn readers, and society in general, so that we can adapt successfully and minimize damaging impacts.

1 THE GREAT BALLOON RACE

Few economists saw our current crisis coming, but this predictive failure was the least of the field's problems. More important was the profession's blindness to the very possibility of catastrophic failures in a market economy.

— Paul Krugman (economist)

The conventional wisdom on the state of the economy — that the financial crisis that started in 2008 was caused by bad real estate loans and that eventually, when the kinks are worked out, the nation will be back to business as usual — is tragically wrong. Our real situation is far more unsettling, our problems have much deeper roots, and an adequate response will require far more from us than just waiting for the business cycle to come back around to the "growth" setting. In reality, our economic system is set for a dramatic, and for all practical purposes permanent, reset to a much lower level of function. Civilization is about to be downsized.

Why have the vast majority of pundits missed this story? Partly because they rely on economic experts with a tunnel vision that ignores the physical limits of planet Earth — the context in which economies operate.

In this chapter we will see in brief outline not only how economies and economic theories have evolved from ancient times to the present, but how and why some modern industrial economies — particularly that of the US — have come to resemble casinos, where a significant proportion of economic activity takes the form of speculative bets on the rise or fall in value of an array of real or illusory assets. And we'll see why all of these developments have led to the fundamental impasse at which we are stuck today.

In order to maximize our perspective, we're going to start our story at the very beginning.

Economic History in Ten Minutes

Throughout over 95 percent of our species' history, we humans lived by hunting and gathering in what anthropologists call *gift economies*.[1] People had no money, and there was neither barter nor trade among members of any given group. Trade did exist, but it occurred only between members of different communities.

It's not hard to see why sharing was the norm within each band of hunter-gatherers, and why trade was restricted to relations with strangers. Groups were small, usually comprising between 15 and 50 persons, and everyone knew and depended on everyone else within the group. Trust was essential to individual survival, and competition would have undermined trust. Trade is an inherently competitive activity: each trader tries to get the best deal possible, even at the expense of other traders. For hunter-gatherers, cooperation — not competition — was the route to success, and so innate competitive drives (especially among males) were moderated through ritual and custom, while a thoroughly entangled condition of mutual indebtedness helped maintain a generally cooperative attitude on everyone's part.

Today we still enjoy vestiges of the gift economy, notably in the family. We don't keep close tabs on how much we are spending on our three-year-old child in an effort to make sure that accounts are settled at some later date; instead, we provide food, shelter, education and more as free gifts, out of love. Yes, parents enjoy psychological rewards, but (at least in the case of mentally healthy parents) there is no conscious process of bargaining, in which we tell the child, "I will give you food and shelter if you repay me with goods and services of equivalent or greater value."

For humans in simple societies, the community was essentially like a family. Freeloading was occasionally a problem, and when it became a drag on the rest of the group it was punished by subtle or not-so-subtle social signals — ultimately, ostracism. But otherwise no one kept score of who owed whom what; to do so would have been considered very bad manners.

We know this from the accounts of 20th-century anthropologists who visited surviving hunter-gatherer societies. Often they reported on the amazing generosity of people who seemed eager to share everything they owned despite having almost no material possessions and being officially listed by aid agencies as among the poorest people on the planet.[2] In some instances anthropologists felt embarrassed by this generosity, and, after being gifted some prized food or a painstakingly hand-made basket, immediately offered a factory-made knife or ornament in return. The anthropologist assumed that natives would be happy to receive the trinkets, but the recipients instead appeared insulted. What had happened? The natives' initial gifts were a way of saying, "You are part of the family; welcome!" But the immediate offering of a gift in return smacked of trade — something only done with strangers. The anthropologist was understood as having said, "No, thanks. I do not wish to be considered part of your family; I want to remain a stranger to you."

By the way, this brief foray into cultural anthropology shouldn't be interpreted as an argument that the hunting-gathering existence represents some ideal of perfection. Partly because simpler societies lacked police and jails, they tended to feature very high levels of interpersonal violence. Accidents were common and average lifespan was short. The gift economy, with both its advantages and limits, was simply a strategy that worked in a certain context, honed by tens of millennia of trial and error.

Here is economic history compressed into one sentence: As societies have grown more complex, larger, more far-flung, and diverse, the tribe-based gift economy has shrunk in importance, while the trade economy has grown to dominate most aspects of people's lives, and has expanded in scope to encompass the entire planet. Is this progress or a process of moral decline? Philosophers have debated the question for centuries. Approve or disapprove, it is what we have done.

With more and more of our daily human interactions based on exchange rather than gifting, we have developed polite ways of being around each other on a daily basis while maintaining an exchange-mediated social distance. This is particularly the case in large cities, where anonymity is fostered also by the practical formalities and psychological impacts that go along with the need to interact with large numbers of strangers, day in and day out. In the best instances, we still take care of one another — often through government programs and private charities. We still enjoy some of the benefits of the old gift economy in our families and churches. But increasingly, the market rules our lives. Our apparent destination in this relentless trajectory toward expansion of trade is a world in which everything is for sale, and all human activities are measured by and for their monetary value.

Humanity has benefited in many obvious ways from this economic evolution: the gift economy really only worked when we lived in small bands and had almost no possessions to speak of. So letting go of the gift economy was a trade-off for houses, cities, cars, iPhones, and all the rest. Still, saying goodbye to community-as-family was painful, and there have been various attempts throughout history to try to revisit it. Communism was one such attempt. However, trying to institutionalize a gift economy at the scale of the nation state introduces all kinds of problems, including those of how to reward initiative and punish laziness in ways that everyone finds acceptable, and how to deter corruption among those whose job it is to collect, count, and reapportion the wealth.

But back to our tour of economic history. Along the road from the gift economy to the trade economy there were several important landmarks. Of these, the invention of money was clearly the most important. Money is a tool used to facilitate trade. People invented it because they needed a medium of exchange to make trading easier, simpler, and more flexible. Once money came into use, the exchange process was freed to grow and to insert itself into aspects of life where it had never been permitted previously. Money simultaneously began to serve other functions as well — principally, as a measure and store of value.

Today we take money for granted. But until fairly recent times it was an oddity, something only merchants used on a daily basis. Some complex societies, including the Inca civilization, managed to do almost completely without it; even in the US, until the mid-20th century, many rural families used money only for occasional trips into town to buy nails, boots, glass, or other items they couldn't grow or make for themselves on the farm.

In his marvelous book *The Structures of Everyday Life: Civilization & Capitalism 15th–18th Century*, historian Fernand Braudel wrote of the gradual insinuation of the money economy into the lives of medieval peasants: "What did it actually bring? Sharp variations in prices of essential foodstuffs; incomprehensible relationships in which man no longer recognized either himself, his customs or his ancient values. His work became a commodity, himself a 'thing.'"[3]

While early forms of money consisted of anything from sheep to shells, coins made of gold and silver gradually emerged as the most practical, universally accepted means of exchange, measure of value, and store of value.

Money's ease of storage enabled industrious individuals to accumulate substantial amounts of wealth. But this concentrated wealth also presented a target for thieves. Thievery was especially a problem for traders: while the portability of money enabled travel over long distances for the purchase of rare fabrics and spices, highwaymen often lurked along the way, ready to snatch a purse at knife-point. These problems led to the invention of banking — a practice in which metal-smiths who routinely dealt with large amounts of gold and silver (and who were accustomed to keeping it in secure, well-guarded vaults) agreed to store other people's coins, offering storage receipts in return. Storage receipts could then be traded as money, thus making trade easier and safer.[4]

Eventually, by the Middle Ages, goldsmith-bankers realized that they could issue these tradable receipts for more gold than they had in their vaults, without anyone being the wiser. They did this by making loans of the receipts, for which they charged a fee amounting to a percentage of the loan.

Initially the church regarded the practice of profiting from loans as a sin — known as *usury* — but the bankers found a loophole in religious doctrine: it was permitted to charge for reimbursement of expenses incurred in making the loan. This was termed *interest*. Gradually bankers widened the definition of "interest" to include what had formerly been called "usury."

The practice of loaning out receipts for gold that didn't really exist worked fine, unless many receipt-holders wanted to redeem paper notes for gold or silver all at once. Fortunately for the bankers, this happened so rarely that eventually the writing of receipts for more money than was on deposit became a perfectly respectable practice known as fractional reserve banking.

It turned out that having increasing amounts of money in circulation was a benefit to traders and industrialists during the historical period when all of this was happening — a time when unprecedented amounts of new wealth were being created, first through colonialism and slavery, but then by harnessing the enormous energies of fossil fuels.

The last impediment to money's ability to act as a lubricant for transactions was its remaining tie to precious metals. As long as paper notes were redeemable for gold or silver, the amounts of these substances existing in vaults put at least a theoretical restraint on the process of money creation. Paper currencies not backed by metal had sprung up from time to time, starting as early as the 13th century CE in China; by the late 20th century, they were the near-universal norm.

Along with more abstract forms of currency, the past century has also seen the appearance and growth of ever more sophisticated investment instruments. Stocks, bonds, options, futures, long- and short-selling, credit default swaps, and more now enable investors to make (or lose) money on the movement of prices of real or imaginary properties and commodities, and to insure their bets — even their bets on other investors' bets.

Probably the most infamous investment scheme of all time was created by Charles Ponzi, an Italian immigrant to the US who, in 1919, began promising investors he could double their money within 90 days. Ponzi told clients the profits would come from buying discounted postal reply coupons in other countries and redeeming them at face value in the United States — a technically legal practice that could yield up to a 400 percent profit on each coupon redeemed due to differences in currency values. What he didn't tell them was that each coupon had to be redeemed individually, so the red tape involved would entail prohibitive costs if large numbers of the coupons (which were only worth a few pennies) were bought and redeemed. In reality, Ponzi was merely paying early investors returns from the principal amounts contributed by later investors. It was a way of shifting wealth from the many to the few, with Ponzi skimming off a lavish income as the money passed through his hands. At the height of the scheme, Ponzi was raking in $250,000 a day, millions in today's dollars. Thousands of people lost their savings, in some cases having mortgaged or sold their houses in order to invest.

A few critics (primarily advocates of gold-backed currency) have called fractional reserve banking a kind of Ponzi scheme, and there is some truth to the claim.[5] As long as the real economy of goods and services within a nation is growing, an expanding money supply seems justifiable, arguably necessary. However, units of currency are essentially claims on labor and natural resources—and as those claims multiply (with the growth of the money supply), and as resources deplete, eventually the remaining resources will be insufficient to satisfy all of the existing monetary claims. Those claims will lose value, perhaps dramatically and suddenly. When this happens, paper and electronic currency systems based on money creation through fractional reserve banking will produce results somewhat similar to those of a collapsing Ponzi scheme: the vast majority of those involved will lose much or all of what they thought they had.

BOX 1.1 Why Was Usury Banned?

In his book *Medici Money: Banking, Metaphysics, and Art in Fifteenth-Century Florence*, Tim Parks writes:

"Usury changes things. With interest rates, money is no longer a simple and stable commodity that just happens to have been chosen as a medium of exchange. Projected through time, it multiplies, and this without any toil on the part of the usurer. Everything becomes more fluid. A man can borrow money, buy a loom, sell his wool at a high price, change his station in life. Another man can borrow money, buy the first man's wool, ship it abroad, and sell it at an even higher price. He moves up the social scale. Or if he is unlucky, or foolish, he is ruined. Meanwhile, the usurer, the banker, grows richer and richer. We can't even know how rich, because money can be moved and hidden, and gains on financial transactions are hard to trace. It's pointless to count his sheep and cattle or to measure how much land he owns. Who will make him pay his tithe? Who will make him pay his taxes? Who will persuade him to pay some attention to his soul when life has become so *interesting*? Things are getting out of hand."[6]

Economics for the Hurried

We have just surveyed the history of *economies*—the systems by which humans create and distribute wealth. *Economics*, in contrast, is a set of philosophies, ideas, equations, and assumptions that describe how all of this does, or should, work.[7]

This story begins much more recently. While the first economists were ancient Greek and Indian philosophers, among them Aristotle (382–322 BCE)—who discussed the "art" of wealth acquisition and questioned whether property should best be owned privately or by government acting on behalf of the people—little of real substance was added to the discussion during the next two thousand years.

It's in the 18th century that economic thinking really gets going. "Classical" economic philosophers such as Adam Smith (1723–1790), Thomas Robert Malthus (1766–1834), and David Ricardo (1772–1823) introduced basic concepts such as supply and demand, division of labor, and the balance of international trade. As happens in so many disciplines, early practitioners were presented with plenty of uncharted territory and proceeded to formulate general maps of their subject that future experts would labor to refine in ever more trivial ways.

These pioneers set out to discover natural laws in the day-to-day workings of economies. They were striving, that is, to make of economics a science on a par with the emerging disciplines of physics and astronomy.

Like all thinkers, the classical economic theorists—to be properly understood—must be viewed in the context of their age. In the 17th and 18th centuries, Europe's power structure was beginning to strain: as wealth flowed from colonies, merchants and traders were getting rich, but they increasingly felt hemmed in by the established privileges of the aristocracy and the church. While economic philosophers were mostly interested in questioning the aristocracy's entrenched advantages, they admired the ability of physicists, biologists, and astronomers to demonstrate the fallacy of old church doctrines, and to establish new universal "laws" through inquiry and experiment.

Physical scientists set aside biblical and Aristotelian doctrines about how the world works and undertook active investigations of natural phenomena such as gravity and electromagnetism—fundamental forces of nature. Economic philosophers, for their part, could point to price as arbiter of supply and demand, acting everywhere to allocate resources far more effectively than any human manager or bureaucrat could ever possibly do. Surely this was a principle as universal and impersonal as the force of gravitation! Isaac Newton had shown there was more to the motions of the stars and planets than could be found in the book of Genesis; similarly, Adam Smith was revealing more potential in the principles and practice of trade than had ever been realized through the ancient, formal relations between princes and peasants, or among members of the medieval crafts guilds.

The classical theorists gradually adopted the math and some of the terminology of science. Unfortunately, however, they were unable to incorporate into economics the basic self-correcting methodology that is science's defining characteristic. Economic theory required no falsifiable hypotheses and demanded no repeatable controlled experiments (these would in most instances have been hard to organize in any case). Economists began to think of themselves as scientists, while in fact their discipline remained a branch of moral philosophy — as it largely does to this day.[8]

The notions of these 18th- and early 19th-century economic philosophers constituted classical economic *liberalism* — the term *liberal* in this case indicating a belief that managers of the economy should let markets act freely and openly, without outside intervention, to set prices and thereby allocate goods, services, and wealth. Hence the term *laissez-faire* (from the French "let do" or "let it be").

In theory, the Market was a beneficent quasi-deity tirelessly working for everyone's good by distributing the bounty of nature and the products of human labor as efficiently and fairly as possible. But in fact everybody wasn't benefiting equally or (in many people's minds) fairly from colonialism and industrialization. The Market worked especially to the advantage of those for whom making money was a primary interest in life (bankers, traders, industrialists, and investors), and who happened to be clever and lucky. It also worked nicely for those who were born rich and who managed not to squander their birthright. Others, who were more interested in growing crops, teaching children, or taking care of the elderly, or who were forced by circumstance to give up farming or cottage industries in favor of factory work, seemed to be getting less and less — certainly as a share of the entire economy, and often in absolute terms. Was this fair? Well, that was a moral and philosophical question. In defense of the Market, many economists said that it *was* fair: merchants and factory owners were making more because they were increasing the general level of economic activity; as a result, everyone else would also benefit…eventually. See? The Market can do no wrong. To some this sounded a bit like the circularly reasoned response of a medieval priest to doubts about the infallibility of scripture. Still, despite its blind spots, classical economics proved useful in making sense of the messy details of money and markets.

Importantly, these early philosophers had some inkling of natural limits and anticipated an eventual end to economic growth. The essential ingredients of the economy were understood to consist of *land, labor*, and *capital*. There was on Earth only so much land (which in these theorists' minds stood for all natural resources), so of course at some point the expansion of the economy would cease. Both Malthus and Smith explicitly held this view. A somewhat later economic philosopher, John Stuart Mill (1806–1873), put the matter as follows: "It must always have been seen, more or less distinctly, by political economists, that the increase in wealth is not boundless: that at the end of what they term the progressive state lies the stationary state…."[9]

But, starting with Adam Smith, the idea that continuous "improvement" in the human condition was possible came to be generally accepted. At first, the meaning of "improvement" (or *progress*) was kept vague, perhaps purposefully. Gradually, however, "improvement" and "progress" came to mean "growth" in the current economic sense of the term — abstractly, an increase in Gross Domestic Product (GDP), but in practical terms, an increase in consumption.

A key to this transformation was the gradual deletion by economists of *land* from the theoretical primary ingredients of the economy (increasingly, only labor and capital really mattered, land having been demoted to a sub-category of capital). This was one of the refinements that turned classical economic theory into *neoclassical* economics; others included the theories of utility maximization and rational choice. While this shift began in the 19th century, it reached its fruition in the 20th through the work of economists who explored models of imperfect competition, and theories of market forms and industrial organization, while emphasizing tools such as the marginal revenue curve (this is when economics came to be known as "the dismal science" — partly because its terminology was, perhaps intentionally, increasingly mind-numbing).[10]

Meanwhile, however, the most influential economist of the 19th century, a philosopher named Karl Marx, had thrown a metaphorical bomb through the window of the house that Adam Smith had built. In his most important book, *Das Kapital*, Marx proposed a name for the economic system that had evolved since the Middle Ages: *capitalism*. It was a system founded on capital. Many people assume that *capital* is simply another word for *money*, but that entirely misses the essential point: capital is wealth — money, land, buildings, or machinery — that has been set aside for production of more wealth. If you use your entire weekly paycheck for rent, groceries, and other necessities, you may occasionally have money but no capital. But even if you are deeply in debt, if you own stocks or bonds, or a computer that you use for a home-based business, you have capital.

Capitalism, as Marx defined it, is a system in which productive wealth is privately owned. *Communism* (which Marx proposed as an alternative) is one in which productive wealth is owned by the community, or by the nation on behalf of the people.

In any case, Marx said, capital tends to grow. If capital is privately held, it *must* grow: as capitalists compete with one another, those who increase their capital fastest are inclined to absorb the capital of others who lag behind, so the system as a whole has a built-in expansionist imperative. Marx also wrote that capitalism is inherently unsustainable, in that when the workers become sufficiently impoverished by the capitalists, they will rise up and overthrow their bosses and establish a communist state (or, eventually, a stateless workers' paradise).

The ruthless capitalism of the 19th century resulted in booms and busts, and a great increase in inequality of wealth — and therefore an increase in social unrest. With the depression of 1873 and the crash of 1907, and finally the Great Depression of the 1930s, it appeared to many social commentators of the time that capitalism was indeed failing, and that Marx-inspired uprisings were inevitable; the Bolshevik revolt in 1917 served as a stark confirmation of those hopes or fears (depending on one's point of view).

20th-Century Economics

Beginning in the late 19th century, social liberalism emerged as a moderate response to both naked capitalism and Marxism. Pioneered by sociologist Lester F. Ward (1841–1913), psychologist William James (1842–1910), philosopher John Dewey (1859–1952), and physician-essayist Oliver Wendell Holmes (1809–1894), social liberalism argued that government has a legitimate economic role in addressing social issues such as unemployment, healthcare, and education. Social liberals decried the unbridled concentration of wealth within society and the conditions suffered by factory workers, while expressing sympathy for labor unions. Their general goal was to retain the dynamism of private capital while curbing its excesses.

Non-Marxian economists channeled social liberalism into economic reforms such as the progressive income tax and restraints on monopolies. The most influential of the early 20th-century economists of this school was John Maynard Keynes (1883–1946), who advised that when the economy falls into a recession government should spend lavishly in order to restart growth. Franklin Roosevelt's New Deal programs of the 1930s constituted a laboratory for Keynesian economics, and the enormous scale of government borrowing and spending during World War II was generally credited with ending the Depression and setting the US on a path of economic expansion.

The next few decades saw a three-way contest between Keynesian social liberals, the followers of Marx, and temporarily marginalized neoclassical or neoliberal economists who insisted that social reforms and government borrowing or meddling with interest rates merely impeded the ultimate efficiency of the free Market.

With the fall of the Soviet Union at the end of the 1980s, Marxism ceased to have much of a credible voice in economics. Its virtual disappearance from the discussion created space for the rapid rise of the neoliberals, who for some time had been drawing energy from widespread reactions against the repression and inefficiencies of state-run economies. Margaret Thatcher and Ronald Reagan both relied heavily on advice from neoliberal thinkers like monetarist Milton Friedman (1912–2006) and followers of the Austrian School economist Friedrich von Hayek (1899–1992).

There is a saying now in Russia: Marx was wrong in everything he said about communism, but he was right in everything he wrote about capitalism. Since the 1980s, the nearly worldwide re-embrace of classical economic philosophy has predictably led to increasing inequalities of wealth within the US and other nations, and to more frequent and severe economic bubbles and crashes.

Which brings us to the global crisis that began in 2007–2008. By this time the two remaining mainstream economics camps—the Keynesians and the neoliberals—had come to assume that perpetual growth is the rational and achievable goal of national economies. The discussion was only about how to maintain it: through government intervention or a laissez-faire approach that assumes the Market always knows best.

But in 2008 economic growth ceased in many nations, and there has as yet been limited success in restarting it. Indeed, by some measures the US economy is slipping further behind, or at best treading water. This dire reality constitutes a conundrum for both economic camps. It is clearly a challenge to the neoliberals, whose deregulatory policies were largely responsible for creating the shadow banking system, the implosion of which is generally credited with stoking the current economic crisis. But it is a problem also for the Keynesians, whose stimulus packages have failed in their aim of increasing employment and general economic activity. What we have, then, is a crisis not just of the economy, but also of economic theory and philosophy.

The ideological clash between Keynesians and neoliberals (represented to a certain degree in the escalating all-out warfare between the US Democratic and Republican political parties) will no doubt continue and even intensify. But the ensuing heat of battle will yield little light if both philosophies conceal the same fundamental errors. One such error is the belief that economies can and should perpetually grow.

But that error rests on another that is deeper and subtler. The subsuming of *land* within the category of *capital* by nearly all post-classical economists had amounted to a declaration that Nature is merely a subset of the human economy—an endless pile of resources to be transformed into wealth. It also meant that natural resources could always be substituted with some other form of capital—money or technology.[11] The reality, of course, is that the human economy exists within and entirely depends upon Nature, and many natural resources have no realistic substitutes. This fundamental logical and philosophical mistake, embedded at the very core of modern mainstream economic philosophies, set society directly on a course toward the current era of climate change and resource depletion, and its persistence makes conventional economic theories—of both Keynesian and neoliberal varieties—utterly incapable of dealing with the economic and environmental survival threats to civilization in the 21st century.

For help we can look to the ecological and biophysical economists—whose ideas we will discuss in Chapter 6, and who have been thoroughly marginalized by the high priests and gatekeepers of mainstream economics—and, to a certain extent, to the likewise marginalized Austrian and Post-Keynesian schools, whose standard bearers have been particularly good at forecasting and diagnosing the purely financial aspects of the current global crisis. But that help will not come in the form that many would wish: as advice that can return our economy to a "normal" state of "healthy" growth. One way or the other—whether through planning and methodical reform, or through collapse and failure—our economy is destined to shrink, not grow.

BOX 1.2 Absurdities of Conventional Economic Theory

• Mainstream economists' way of calculating a nation's economic health—the Gross Domestic Product (GDP)—counts only monetary transactions. If a country has happy families, the GDP won't reflect that fact; but if the same country suffers a war or natural disaster monetary transactions will likely increase, leading to a bounce in the GDP. Calculating a nation's overall health according to its GDP makes about as much sense as evaluating the quality of a piece of music solely by counting the number of notes it contains.

• A related absurdity is what economists call an "externality." An externality occurs when production or consumption by one party directly affects the welfare of another party, where "directly" means that the effect is unpriced (it is *external* to the market). The damage to ecosystems that occurs from logging and mining is an externality if it isn't figured into the price of lumber or coal. Positive externalities are possible (if some people farm organically, even people who don't grow or eat organic food will benefit thanks to an overall reduction in the load of pesticides in the environment). Unfortunately, negative externalities are far more prevalent, since corporations use them as economic loopholes through which to pump every imaginable sort of pollution and abuse. Corporations keep the profit and leave society as a whole to clean up the mess.

• Mainstream economists habitually treat asset depletion as income, while ignoring the value of the assets themselves. If the owner of an old-growth forest cuts it and sells the timber, the market may record a drop in the land's monetary value, but otherwise the ecological damage done is regarded as an externality. Irreplaceable biological assets, in this case, have been liquidated; thus the benefit of these assets to future generations is denied. From an ecosystem point of view, an economy that does not heavily tax the extraction of non-renewable resources is like a jobless person rapidly spending an inheritance.

• Mainstream economists like to treat people as if they were producers and consumers—and nothing more. The theoretical entity *Homo economicus* will act rationally to acquire as much wealth as possible and to consume as much stuff as possible. Generosity and self-limitation are (according to theory) irrational. Anthropological evidence of the existence of non-economic motives in humans is simply brushed aside. Unfortunately, people tend to act (to some degree, at least) the way they are expected and conditioned to act; thus *Homo economicus* becomes a self-confirming prediction.

Business Cycles, Interest Rates, and Central Banks

We have just reviewed a minimalist history of human economies and the economic theories that have been invented to explain and manage them. But there is a lot of detail to be filled in if we are to understand what's happening in the world economy *today*. And much of that detail has to do with the spectacular growth of debt — in obvious and subtle forms — that has occurred during the past few decades. The modern debt phenomenon in turn must be seen in light of recurring *business cycles* that characterize economic activity in modern industrial societies, and the central banks that have been set up to manage them.

We've already noted that nations learned to support the fossil fuel-stoked growth of their real economies by increasing their money supply via fractional reserve banking. As money was gradually de-linked from physical substance (i.e., precious metals), the creation of money became tied to the making of loans by commercial banks. This meant that the supply of money was entirely elastic — as much could be created as was needed, and the amount in circulation could contract as well as expand. Growth of money was tied to growth of debt.

The system is dynamic and unstable, and this instability manifests in the business cycle, which in a simplified model looks something like this.[12] In the expansionary phase of the cycle, businesses see the future as rosy, and therefore take out loans to build more productive capacity and hire new workers. Because many businesses are doing this at the same time, the pool of available workers shrinks; so, to attract and keep the best workers, businesses have to raise wages. With wages increasing, worker-consumers have more money in their pockets, which they then spend on products made by businesses. This increases demand, and businesses see the future as even rosier and take out more loans to build even more productive capacity and hire even more workers...and so the cycle continues. Amid all this euphoria, workers go into debt based on the expectation that their wages will continue to grow, making it easy to repay loans. Businesses go into debt to expand their productive capacity. Real estate prices go up because of rising demand (former renters decide they can now afford to buy), which means that houses are worth more as collateral for homeowner loans. All of this borrowing and spending increases both the money supply and the "velocity" of money — the rate at which it is spent and re-spent.

At some point, however, the overall mood of the country changes. Businesses have invested in as much productive capacity as they are likely to need for a while. They feel they have taken on as much debt as they can handle and don't need to hire more employees. Upward pressure on wages ceases, which helps dampen the general sense of optimism about the economy. Workers likewise become shy about taking on more debt, and instead concentrate on paying off existing debts. Or, in the worst case, if they have lost their jobs, they may fail to make debt payments or even declare bankruptcy. With fewer loans being written, less new money is being created; meanwhile, as earlier loans are paid off or defaulted upon, money effectively disappears from the system. The nation's money supply contracts in a self-reinforcing spiral.

But if people increase their savings during this downward segment of the cycle, they eventually will feel more secure and therefore more willing to begin spending again. Also, businesses will eventually have liquidated much of their surplus productive capacity and reduced their debt burden. This sets the stage for the next expansion phase.

Business cycles can be gentle or rough, and their timing is somewhat random and largely unpredictable.[13] They are also controversial: Austrian School and Chicago School economists believe they are self-correcting as long as the government and central banks (which we'll discuss below) don't interfere; Keynesians believe they are only partially self-correcting and must be managed.

In the worst case, the upside of the cycle can constitute a bubble, and the downside a recession or even a depression. A *recession* is a widespread decline in GDP, employment, and trade lasting from six months to a year; a *depression* is a sustained, multi-year contraction in economic activity. In the narrow sense of the term, a *bubble* consists of trade in high volumes at prices that are considerably at odds with intrinsic values, but the word can also be used more broadly to refer to any instance of rapid expansion of currency or credit that's not sustainable over the long run. Bubbles always end with a crash: a rapid, sharp decline in asset values.

Interest rates can play an important role in the business cycle. When rates are low, both businesses and individuals are more likely to want to take on more debt; when rates are high, new debt is more expensive to service. When money is flooding the system, the price of money (in terms of interest rates) naturally tends to fall, and when money is tight its price tends to rise — effects that magnify the existing trend.[14]

During the 19th century, as banks acted with little supervision in creating money to fuel business growth cycles and bubbles, a series of financial crises ensued. In response, bankers in many countries organized to pressure governments to authorize central banks to manage the national money supply. In the US, the Federal Reserve ("the Fed") was authorized by Congress in 1913 to act as the nation's central bank.

The essential role of central banks, such as the Fed, is to conduct the nation's monetary policy, supervise and regulate banks, maintain the stability of the financial system, and provide financial services to both banks and the government. In doing this, central banks also often aim to moderate business cycles by controlling interest rates. The idea is simple enough: lowering interest rates makes borrowing easier, leading to an increasing money supply and the moderation of recessionary trends; high interest rates discourage borrowing and deflate dangerous bubbles.

The Federal Reserve charters member banks, which must obey rules if they are to maintain the privilege of creating money through generating loans. It effectively controls interest rates for the banking system as a whole by influencing the rate that banks charge each other for overnight loans of federal funds, and the rate for overnight loans that member banks borrow directly from the Fed. In addition, the Fed can purchase government debt obligations, creating the money out of thin air (by *fiat*) with which to do so, thus directly expanding the nation's money supply and thereby influencing the interest rates on bonds.

The Fed has often been a magnet for controversy. While it operates without fanfare and issues statements filled with terms opaque even to many trained economists, its secrecy and power have led many critics to call for reforms or for its replacement with other kinds of banking regulatory institutions. Critics point out that the Fed is not really democratic (the Fed chairman is appointed by the US President, but other board members are chosen by private banks, which also own shares in the institution, making it an odd government-corporate hybrid).

Other central banks serve similar functions within their domestic economies, but with some differences: The Bank of England, for example, was nationalized in 1946 and is now wholly owned by the government; the Bank of Russia was set up in 1990 and by law must channel half of its profits into the national budget (the Fed does this with all its profits, after deducting operating expenses). Nevertheless, many see the Fed and central banks elsewhere (the European Central Bank, the Bank of Canada, the People's Bank of China, the Reserve Bank of India) as clubs of bankers that run national economies largely for their own benefit. Suspicions are most often voiced with regard to the Fed itself, which is arguably the most secretive and certainly the most powerful of the central banks. Consider the Fed's theoretical ability to engineer either a euphoric financial bubble or a Wall Street crash immediately before an election, and its ability therefore to substantially impact that election. It is not hard to see why president James Garfield would write, "Whoever controls the volume of money in any country is absolute master of industry and commerce," or why Thomas Jefferson would opine, "Banking establishments are more dangerous than standing armies."

Still, the US government itself—apart from the Fed—maintains an enormous role in managing the economy. National governments set and collect taxes, which encourage or discourage various kinds of economic activity (taxes on cigarettes encourage smokers to quit; tax breaks for oil companies discourage alternative energy producers). General tax cuts can spur more activity throughout the economy, while generally higher taxes may dampen borrowing and spending. Governments also regulate the financial system by setting their own rules for banks, insurance companies, and investment institutions.

Meanwhile, as Keynes advised, governments also borrow and spend to create infrastructure and jobs, becoming the borrowers and spenders of last resort during recessions. A non-trivial example: In the US since World War II, military spending has supported a substantial segment of the national economy—the weapons industries and various private military contractors—while directly providing hundreds of thousands of jobs, at any given moment, for soldiers and support personnel. Critics describe the system as a military-industrial "welfare state for corporations."[15]

The upsides and downsides of the business cycle are reflected in higher or lower levels of inflation. Inflation is often defined in terms of higher wages and prices, but (as Austrian School economists have persuasively argued) wage and price inflation is actually just the symptom of an increase in the money supply relative to the amounts of goods and services being traded, which in turn is typically the result of exuberant borrowing and spending. Inflation causes each unit of currency to lose value. The downside of the business cycle, in the worst instance, can produce the opposite of inflation, or *deflation*. Deflation manifests as declining wages and prices, due to a declining money supply relative to goods and services traded (which causes each unit of currency to increase in purchasing power), itself due to a contraction of borrowing and spending or to widespread defaults.

Business cycles, and regulated monetary and banking systems, constitute the framework within which companies, investors, workers, and consumers act. But over the past few decades something remarkable has happened within that framework. In the US, the financial services industry has ballooned to unprecedented proportions, and has plunged society as a whole into a crisis of still-unknown proportions. How and why did this happen? As we are about to see, these recent developments have deep roots.

Mad Money

Investing is a practice nearly as old as money itself, and from the earliest times motives for investment were two-fold: to share in profits from productive enterprise, and to speculate on anticipated growth in the value of assets. The former kind of investment is generally regarded as helpful to society, while the latter is seen, by some at least, as a form of gambling that eventually results in wasteful destruction of wealth. It is important to remember that the difference between the two is not always clear-cut, as investment always carries risk as well as an expectation of reward.[16]

Here are obvious examples of the two kinds of investment motive. If you own shares of stock in General Motors, you own part of the company; if it does well, you are paid dividends — in "normal" times, a modest but steady return on your investment. If dividends are your main objective, you are likely to hold your GM stock for a long time, and if most others who own GM stock have bought it with similar goals, then — barring serious mismanagement or a general economic downturn — the value of the stock is likely to remain fairly stable. But suppose instead you bought shares of a small start-up company that is working to perfect a new oil-drilling technology. If the technology works, the value of the shares could skyrocket long before the company actually shows a profit. You could then dump your shares and make a killing. If you're this kind of investor, you are more likely to hold shares relatively briefly, and you are likely to gravitate toward stocks that see rapid swings in value. You are also likely to be constantly on the lookout for information — even rumors — that could tip you off to impending price swings in particular stocks.

When lots of people engage in speculative investment, the likely result is a series of occasional manias or bubbles. A classic example is the 17th-century Dutch tulip mania, when trade in tulip bulbs assumed bubble proportions; at its peak in early February 1637, some single tulip bulbs sold for more than ten times the annual income of a skilled craftsman.[17] Just days after the peak, tulip bulb contract prices collapsed and speculative tulip trading virtually ceased. More recently, in the 1920s, radio stocks were the bubble *du jour* while the dot-com or Internet bubble ran its course a little over a decade ago (1995–2000).

Given the evident fact that bubbles tend to burst, resulting in a destruction of wealth sometimes on an enormous and catastrophic scale, one might expect that governments would seek to restrain the riskier versions of speculative investing through regulation. This has indeed been the case in historic periods immediately following spectacular crashes. For example, after the 1929 stock market crash regular commercial banks (which accept deposits and make loans) were prohibited from acting as investment banks (which deal in stocks, bonds, and other financial instruments). But as the memory of a crash fades, such restraints tend to fall away.

Moreover, investors are always looking for creative ways to turn a profit — sometimes by devising new methods that are not yet constrained by regulations. A few of these methods were particularly instrumental in the build-up to the 2007–2008 crisis. As we discuss them, we will also define some crucial terms.

Let's start with *leverage* — a general term for any way to multiply investment gains or losses. A bit of history helps in understanding the concept. During the 1920s, partly because the Fed was keeping interest rates low, investors found they could borrow money to buy stocks, then make enough of a profit in the buoyant stock market to repay their debt (with interest) and still come out ahead. This was called *buying on margin*, and it is a classic form of leverage. Unfortunately, when worries about higher interest rates and falling real estate prices helped trigger the stock market crash of October 1929, margin investors found themselves owing enormous sums they couldn't repay. The lesson: leverage can multiply profits, but it likewise multiplies losses.[18]

Two important ways to attain leverage are by borrowing money and trading securities. An example of the former: A public corporation (i.e., one that sells stock) may leverage its equity by borrowing money. The more it borrows, the fewer dividend-paying stock shares it needs to sell to raise capital, so any profits or losses are divided among a smaller base and are proportionately larger as a result. The company's stock looks like a better buy and the value of shares may increase. But if a corporation borrows too much money, a business downturn might drive it into bankruptcy, while a less-leveraged corporation might prove more resilient.

In the financial world, leverage is mostly achieved with securities. A *security* is any fungible, negotiable financial instrument representing value. Securities are generally categorized as debt securities (such as bonds and debentures), equity securities (such as common stocks), and derivative contracts.

Debt and equity securities are relatively easy to explain and understand; derivatives are another story. A *derivative* is an agreement between two parties that has a value that is determined by the price movement of something else (called the *underlying*). The underlying can consist of stock shares, a currency, or an interest rate, to cite three common examples. Since a derivative can be placed on any sort of security, the scope of possible derivatives is nearly endless. Derivatives can be used either to deliberately acquire risk (and increase potential profits) or to hedge against risk (and reduce potential losses). The most widespread kinds of derivatives are *options* (financial instruments that give owners the right, but not the obligation, to engage in a specific transaction on an asset), *futures* (a contract to buy or sell an asset at a future date at a price agreed today), and *swaps* (in which counterparties exchange certain benefits of

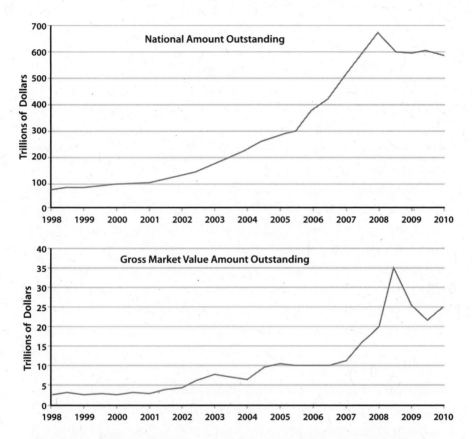

FIGURE 9. Amounts Outstanding of Over the Counter (OTC) Derivatives since 1998 in G10 Countries and Switzerland. "Notional value" refers to the total value of a leveraged position's assets. The term is commonly used in the options, futures, and currency markets when a small amount of invested money controls a large position (and has a large consequence for the trader). "Market value" refers to how much derivatives contracts would be worth if they had to be settled at a given moment. Source: Bank for International Settlements.

one party's financial instrument for those of the other party's financial instrument).

Derivatives have a fairly long history: rice futures have been traded on the Dojima Rice Exchange in Osaka, Japan since 1710. However, they have more recently attracted considerable controversy, as the total nominal value of outstanding derivatives contracts has grown to colossal proportions — in the hundreds of trillions of dollars globally, according to some estimates. Prior to the crash of 2008, investor Warren Buffett famously called derivatives "financial weapons of mass destruction," and asserted that they constitute an enormous bubble. Indeed, during the 2008 crash, a subsidiary of the giant insurance company AIG lost more than $18 billion on a type of swap known as a *credit default swap*, or CDS (essentially an insurance arrangement in which the buyer pays a premium at periodic intervals in exchange for a contingent payment in the event that a third party defaults). Société Générale lost $7.2 billion in January of the same year on futures contracts.

Often, mundane financial jargon conceals truly remarkable practices. Take the common terms *long* and *short* for example. If a trader is "long" on oil futures, for example, that means he or she is holding contracts to buy or sell a specified amount of oil at a specified future date at a price agreed today, in expectation of a rise in price. One would therefore naturally assume that taking a "short" position on oil futures or anything else would involve expectation of a falling price. True enough. But just how does one successfully go about investing to profit on assets whose value is declining? The answer: *short selling* (also known as *shorting* or *going short*), which involves borrowing the assets (usually securities borrowed from a broker, for a fee) and immediately selling them, waiting for the price of those assets to fall, buying them back at the depressed price, then returning them to the lender and pocketing the price difference. Of course, if the price of the assets rises, the short seller loses money. If this sounds dodgy, then consider *naked short selling*, in which the investor sells a financial instrument without bothering first to buy or borrow it, or even to ensure that it *can* be borrowed. Naked short selling is illegal in the US, but many knowledgeable commentators assert that the practice is widespread nonetheless.

In the boom years leading up to the 2007–2008 crash, it was often the wealthiest individuals who engaged in the riskiest financial behavior. And the wealthy seemed to flock, like finches around a bird feeder, toward *hedge funds*: investment funds that are open to a limited range of investors and that undertake a wider range of activities than traditional "long-only" funds invested in stocks and bonds — activities including short selling and entering into derivative contracts. To neutralize the effect of overall market movement, hedge fund managers balance portfolios by buying assets whose price is expected to outpace the market, and by short-selling assets expected to do worse than the market as a whole. Thus, in theory, price movements of particular securities that reflect overall market activity are cancelled out, or "hedged." Hedge funds promise (and often produce) high returns through extreme leverage. But because of the enormous sums at stake, critics say this poses a systemic risk to the entire economy. This risk was highlighted by the near-collapse of two Bear Stearns hedge funds, which had invested heavily in mortgage-backed securities, in June 2007.[19]

I Owe You

As we have seen, bubbles are a phenomenon generally tied to speculative investing. But in a larger sense our entire economy has assumed the characteristics of a bubble or a Ponzi scheme. That is because it has come to depend upon staggering and continually expanding amounts of debt: government and private debt; debt in the trillions, and tens of trillions, and hundreds of trillions of dollars; debt that, in aggregate, has grown by 500 percent since 1980; debt that has grown faster than economic output (measured in GDP) in all but one of the past 50 years; debt that can never be repaid; debt that represents claims on quantities of labor and resources that simply do not exist.

When we inquire how and why this happened, we discover a web of interrelated trends.

Looking at the problem close up, the globalization of the economy looms as a prominent factor. In the 1970s and '80s, with stiffer environmental and labor standards to contend with domestically, corporations began eyeing the regulatory vacuum, cheap labor, and relatively untouched natural resources of less-industrialized nations as a potential goldmine. International investment banks started loaning poor nations enormous sums to pay for ill-advised infrastructure projects (and, incidentally, to pay kickbacks to corrupt local politicians), later requiring these countries to liquidate their natural resources at fire-sale prices so as to come up with the cash required to make loan payments. Then, prodded by corporate interests, industrialized nations pressed for the liberalization of trade rules via the World Trade Organization (the new rules almost always subtly favored the wealthier trading partner). All of this led predictably to a reduction of manufacturing and resource extraction in core industrial nations, especially the US (many important resources were becoming depleted in the wealthy industrial nations anyway), and a steep increase in resource extraction and manufacturing in several "developing" nations, principally China. Reductions in domestic manufacturing and resource extraction in turn motivated investors within industrial nations to seek profits through purely financial means. As a result of these trends, there are now as many Americans employed in manufacturing as there were in 1940, when the nation's population was roughly half what it is today, while the proportion of total US economic activity deriving from financial services has tripled during the same period. Speculative investing has become an accepted practice that is taught in top universities and institutionalized in the world's largest corporations.

But as we back up to take in a wider view, we notice larger and longer-term trends that have played even more important roles. One key factor was the severance of money from its moorings in precious metals, a process that started over a century ago. Once money came to be based on debt (so that it was created primarily when banks made loans), growth in total outstanding debt became a precondition for growth of the money supply and therefore for economic expansion. With virtually everyone — workers, investors, politicians — clamoring for more economic growth, it was inevitable that innovative ways to stimulate the process of debt creation would be found. Hence the fairly recent appearance of a bewildering array of devices for borrowing, betting, and insuring — from credit cards to credit default swaps — all essentially tools for the "ephemeralization" of money and the expansion of debt.

A Marxist would say that all of this flows from the inherent imperatives of capitalism. A historian might contend it reflects the inevitable trajectory of all empires (though past empires didn't have fossil fuels and therefore lacked the means to become global in extent). And a cultural anthropologist might point out that the causes of our debt spiral are endemic to civilization itself: as the gift economy shrank and trade grew, the infinitely various strands of mutual obligation that bind together every human community became translated into financial debt (and, as hunter-gatherers intuitively understood, debts within the community can never fully be repaid — nor should they be; and certainly not with interest).

In the end perhaps the modern world's dilemma is as simple as "What goes up must come down." But as we experience the events comprising ascent and decline close up and first-hand, matters don't appear simple at all. We suffer from media bombardment; we're soaked daily in unfiltered and unorganized data; we are blindingly, numbingly overwhelmed by the rapidity of change. But if we are to respond and adapt successfully to all this change, we must have a way of understanding why it is happening, where it might be headed, and what we can do to achieve an optimal outcome under the circumstances. If we are to get it right, we must see both the forest (the big, long-term trends) and the trees (the immediate challenges ahead).

Which brings us to a key question: If the financial economy cannot continue to grow by piling up more debt, then what will happen next?

BOX 1.3   The Magic of Compound Interest

Suppose you have $100. You decide to put it into a savings account that pays you 5 percent interest. After the first year, you have $105. You leave the entire amount in the bank, so at the end of the second year you are collecting 5 percent interest not on $100, but on $105 — which works out to $5.25. So now you have $110.25 in your account. At first this may not seem all that remarkable. But just wait. After three years you have $115.76, then $121.55, then $127.63, then $134.01. After ten years you would have $162.88, and at the end of fourteen years you would have nearly doubled your initial investment. After 29 years you would have about $400, and if you could manage to leave your investment untouched for forty-three years you would have nearly $800. After eighty-six years your heirs could collect $3,200, and after a full century had passed your initial $100 deposit would have grown to nearly $6,200. Of course, if this were a debt rather than an investment, interest would compound similarly.

Somehow these claims on real wealth (goods and services) have multiplied, while the world's stores of natural resources have in many cases actually declined due to depletion of fossil fuels and minerals, or the over-harvesting of forests and fish. Money, if invested or loaned, has the "right" to increase, while nature enjoys no such imperative.

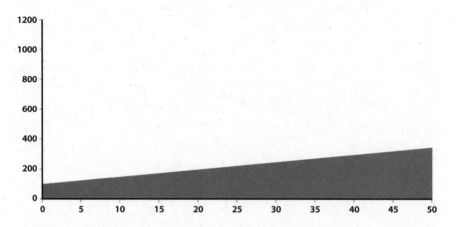

FIGURE 10. Additive growth. Here we see an additive growth rate of 5. Beginning with 100, we add 5, and then add 5 to that sum, and so on. After 50 transactions we arrive at 350.

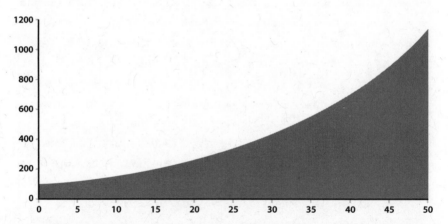

FIGURE 11. Compounded growth. This graph shows a compound growth rate of 5 percent, which means we start by multiplying 100 by 5 percent and then add the product to the original 100. Then we multiply that sum by 5 percent and add the product to the original sum, and so on. After 50 transactions, we arrive at 1147.

CHAPTER 2 THE SOUND OF AIR ESCAPING

We're in the midst of a once-in-a-lifetime set of economic conditions. The perspective I would bring is not one of recession. Rather, the economy is resetting to a lower level of business and consumer spending based largely on the reduced leverage in the economy.

— Steven Ballmer (Chairman, Microsoft Corp.)

If the previous chapter had been written as a novel, one wouldn't have to read long before concluding that it is a story unlikely to end well. But it is not just a story, it is a description of the system in which our lives and the lives of everyone we care about are all embedded. How economic events unfold from here on is a matter of more than idle curiosity or academic interest.

It's not hard to find plenty of opinions about where the economy is, or should be, headed. There are Chicago School economists, who see debt and meddling by government and central banks as problems and austerity as the solution; and Keynesians, who see the problem as being insufficient government stimulus to counter deflationary trends in the economy. There are those who say that bloated government borrowing and spending mean we are in for a currency-killing bout of hyperinflation, and those who say that government cannot inject enough new money into the economy to make up for commercial banks' hesitancy to lend, so the necessary result will be years of deflationary depression. As we'll see, each of these perspectives is probably correct up to a point. Our purpose in this chapter will not be to forecast exactly how the global economic system will behave in the near future—which is impossible in any case because there are too many variables at play—but to offer a brief but fairly comprehensive, non-partisan survey of the factors and forces at work in the post-2008 global financial economy, integrating various points of view as much as possible.

To do this, we will start with a brief overview of the meltdown that began in 2007, then look at the theoretical and practical limits to debt; we will then review the bailout and stimulus packages deployed to lessen the impact of the crisis; and finally we will explore a few scenarios for the short- and mid-term future.

This won't hurt much. Honest.

Houses of Cards

Lakes of printer's ink have been spilled in recounting the events leading up to the financial crisis that began in 2007–2008; here we will add only a small puddle. Nearly everyone agrees that it unfolded in essentially the following steps:

- In an attempt to limit the consequences of the "dot-com" crash of 2000, the Federal Reserve drastically lowered interest rates, enabling lenders across the country to provide *easy credit* to households and businesses who hadn't been able to access it before.
- This led to a *housing bubble*, which was made much worse by *sub-prime lending*.
- Partly because of the prior *deregulation* of the financial industry, the housing bubble was also magnified by *over-leveraging* within the financial services industry, which was in turn exacerbated by *financial innovation and complexity* (including the use of derivatives, collateralized debt obligations, and a dizzying variety of related investment instruments)—all feeding the boom of a *shadow banking system*, whose potential problems were hidden by *incorrect pricing of risk* by ratings agencies.
- A *commodities boom* (which drove up gasoline and food prices) and

temporarily *rising interest rates* (especially on adjustable-rate mortgages) ultimately undermined consumer spending and confidence, helping to burst the housing bubble — which, once it started to deflate, set in motion a chain reaction of defaults and bankruptcies.

Each element of that brief description has been unpacked at great length in books like Andrew Ross Sorkin's *Too Big to Fail* and Bethany McLean's and Joe Nocera's *All the Devils Are Here*, and in the documentary film "Inside Job."[1] It's old, sad news now, though many parts of the story are still controversial (e.g., was the problem *deregulation* or *bad regulation*?). And yet, many analyses overlook the fact that these events were manifestations of a deeper trend toward dramatically and unsustainably increasing debt, credit, and leverage. So it's important that we review this recent history in a little more detail so we can see why, from a purely financial point of view, growth is currently on hold and is unlikely to return for the foreseeable future.

Setting the Stage: 1970 to 2001

Starting in the 1970s, GDP growth rates in Western countries began to taper off. The US had been the world's primary oil producer, but in 1971 its oil production peaked and began to decline. That meant US oil imports would have to increase to compensate, thus encouraging trade deficits. Moreover, domestic markets for major consumer goods were becoming saturated.

In the US, inflation-adjusted wages — particularly for the hourly workers who comprise 80 percent of the workforce — were stagnating after fifteen decades of major gains. Relatively constant wage levels meant that most households couldn't afford to *increase* their spending (remember: the health of the economy requires *growth*) unless they saved less and borrowed more. Which they began to do.

With the rate of growth of the real economy stalling somewhat, profitable investment opportunities in manufacturing companies dwindled; this created a surplus of investment capital looking for high yields. The solution hit upon by wealthy investors was to direct this surplus toward financial markets.

The most important financial development during the 1970s was the growth of *securitization*—the financial practice of pooling various types of contractual debt (such as mortgages, auto loans, or credit card debt) and selling it to investors in the forms of bonds or collateralized mortgage obligations (CMOs). The principal and interest on the debts underlying the security were paid back to investors regularly, while the security itself could be sold and re-sold. Securitization provided an avenue for more investors to fund more debt. In effect, securitization allowed claims on wealth to increase far above previous levels. In the US, aggregate debt began rising faster than GDP, with the debt-to-GDP ratio growing from about 150 percent (where it had been for many years until 1980) up to its current level of about 300 percent. *In fact, US aggregate debt has increased more than GDP for every year since 1965.*

Also starting in the 1970s, economists and policy makers began arguing that, in order to end persistent "stagflation," largely caused by high oil prices, government should cut taxes on the rich—who, seeing more money in their bank accounts, would naturally invest their capital in ways that would ultimately benefit everyone.[2] At the same time, policy makers decided it was time to liberate the financial sector from various New Deal-era restraints so that it could create still more innovative investment opportunities.

Some commentators insist that the Community Reinvestment Act of 1977 (since updated nine times)—which was designed to encourage commercial banks and savings associations to meet the needs of borrowers in low- and moderate-income neighborhoods, and to reduce "redlining"—would later contribute to the housing bubble of 2000–2006. This notion has been widely contested. Nevertheless, the chartering by Congress of mortgage corporations Fannie Mae and Freddie Mac in 1968 and 1970 would certainly have implications much later, when the real estate market crashed in 2007.[3] But we are getting ahead of ourselves.

The process of deregulation and regulatory change continued for the next quarter-century. It included, for example, the Commodity Futures Modernization Act, drafted by Senate Republican Phil Gramm and signed into law by Democratic President Bill Clinton in 2000, which legalized the trafficking in packages of dubious home mortgages.

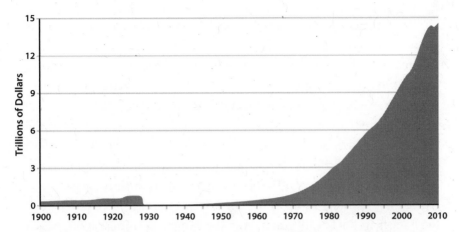

FIGURE 12. US GDP, 1900–2010. This chart shows nominal US Gross Domestic Product (GDP). GDP plummets in 1929 as a result of the stock market crash, then takes almost forty years to recover. We also see the rapid growth of US GDP beginning in 1975 and continuing until the financial crisis of 2008, where we observe a dip in the graph. Source: Years 1900–1928, Louis Johnston and Samuel H. Williamson, "What Was the US GDP Then?" MeasuringWorth, 2010. Years 1929–2010, US Bureau of Economic Analysis.

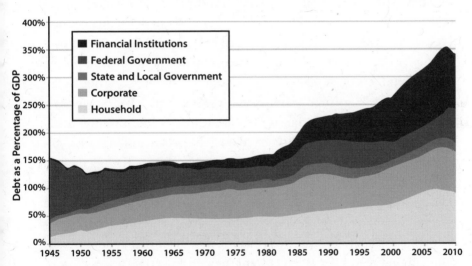

FIGURE 13. Total US Debt as a Percentage of GDP, 1945–2010. Percentages are based on nominal values of both debt and GDP for each year. Government debt (local, state, and federal) remains fairly constant as a percentage of GDP. It is household and financial sector debt that make the largest gains since 1945. We can see financial institutions begin to take on huge levels of debt beginning in the late 1980s, reaching a peak just before the crash of 2008. Source: The Federal Reserve, US Bureau of Economic Analysis.

These regulatory changes were accompanied by a shift in corporate culture: executives began running companies more for the benefit of management than for shareholders, paying themselves spectacular bonuses and putting increasing emphasis on boosting share prices rather than dividends. Auditors, boards of directors, and Wall Street analysts encouraged these trends, convinced that soaring share prices and other financial returns justified them.[4]

America's distribution of income, which had been reasonably equitable during the post-WWII era, began to return to the disparity seen in the 1920s in the lead-up to the Great Depression. This was partly due to changes in tax law, begun during the Reagan administration, which reduced taxes on the wealthiest Americans. In 1970 the top 100 CEOs earned about $45 for every dollar earned by the average worker; by 2008 the ratio was over 1,000 to one.

In the 1990s, as the surplus of financial capital continued to grow, investment banks began inventing a slew of new securities with high yields (and high risk). In assessing these new products, ratings agencies used mathematical models that, in retrospect, seriously underestimated their levels of risk. Decades earlier, bond credit ratings agencies had been paid for their work by investors who wanted impartial information on the credit worthiness of securities issuers and their offerings. Starting in the early 1970s, the "Big Three" ratings agencies (Standard & Poor's, Moody's, and Fitch) began to be paid instead by securities issuers. This eventually led to ratings agencies actively encouraging the issuance of high-risk collateralized debt obligations (CDOs).

Also in the 1990s, the Clinton administration adopted "affordable housing" as one of its explicit goals (this didn't mean lowering house prices; it meant helping Americans get into debt), and over the next decade the percentage of Americans owning their homes increased 7.8 percent. This initiated a persistent upward trend in real estate prices.

The Internet as we know it today opened for business in the mid-1990s, and within a few years investors had bid up Internet-related stocks, creating a speculative bubble. The dot-com bubble burst in 2000 (as with all bubbles, it was only a matter of "when," not "if"), and a year later the terrifying crimes of September 11, 2001 resulted in a four-day closure of US stock exchanges and history's largest one-day decline in the Dow Jones Industrial Average. These events together triggered a significant recession. Seeking to counter the resulting deflationary trend, the Federal Reserve sought to bring interest rates down so as to make borrowing more affordable.

Downward pressure on interest rates was also coming from the nation's high and rising trade deficit. Every nation's balance of payments must sum to zero, so if a nation is running a current account deficit it must balance that amount with funds earned from foreign investments, or by running down reserves, or by obtaining loans from other countries. In other words, a country that imports more than it exports must borrow to pay for those imports. Hence American imports had to be offset by large and growing amounts of foreign investment capital flowing into the US. Higher bond yields attract more investment capital, but there is an inevitable inverse relationship between bond prices and interest rates, so trade deficits tend to force interest rates down.

Foreign investors had plenty of funds to lend, either because they had very high personal savings rates (in China, up to 40 percent of income is saved), or because of high oil prices (a windfall for oil-producing nations). A torrent of funds — a "Giant Pool of Money" doubling in size between 2000 and 2007 — was flowing into US financial markets.[5] While foreign governments were purchasing risk-free US Treasury bonds, thus avoiding much of the impact of the eventual crash, other overseas investors, including pension funds, were gorging on the higher yielding mortgage-backed securities (MBSs) and CDOs. The indirect consequences were that US households were in effect using funds borrowed from foreigners to finance consumption or to bid up house prices, while sales of mortgage-backed securities also amounted to sales of accumulated wealth to foreign investors.

Shadow Banks and the Housing Bubble

By this time a largely unregulated "shadow banking system," made up of hedge funds, money market funds, investment banks, pension funds, and other lightly-regulated entities, had become critical to the credit markets and was underpinning the financial system as a whole. But the shadow "banks" tended to borrow short-term in liquid markets to purchase long-term, illiquid, and risky assets, profiting on the difference between lower short-term rates and higher long-term rates. This meant that any disruption in credit markets would result in rapid deleveraging, forcing these entities to sell long-term assets (such as mortgage-backed securities) at depressed prices.

Between 1997 and 2006, the price of the typical American house increased by 124 percent. House prices were rising much faster than income was growing. During the two decades ending in 2001, the national median home price ranged between 2.9 and 3.1 times median household income. This ratio rose to 4.0 in 2004, and 4.6 in 2006. This meant that, in increasing numbers of cases, people could not actually afford the homes they were buying. Meanwhile, with interest rates low, many homeowners were refinancing their homes, or taking out second mortgages secured by price appreciation, in order to pay for new cars or home remodeling. Many of the mortgages had initially negligible—but adjustable—interest rates, which meant that borrowers would soon face a nasty surprise.

Wall Street had connected the "Giant Pool of Money" to the US mortgage market, with enormous fees accruing throughout the financial supply chain, from the mortgage brokers selling the loans, to small banks funding the brokers, to giant investment banks that would ultimately securitize, bundle, and sell the loans to investors the world over. This capital flow also provided jobs for millions of people in the home construction and real estate industries.

Wall Street brokers began thinking of themselves as each deserving many millions of dollars a year in compensation, simply because they were smart enough to figure out how to send the debt system into overdrive and skim off a tidy percentage for themselves. Bad behavior was being handsomely rewarded, so nearly everyone on Wall Street decided to behave badly.

By around 2003, the supply of mortgages originating under traditional lending standards had largely been exhausted. But demand for MBSs continued, and this helped drive down lending standards—to the point that some adjustable-rate mortgage (ARM) loans were being offered at no initial interest, or with no down payment, or to borrowers with no evidence of ability to pay, or all of the above.

Bundled into MBSs, sold to pension funds and investment banks, and hedged with derivatives contracts, mortgage debt became the very fabric of the US financial system, and, increasingly, the economies of many other nations as well. By 2005 mortgage-related activities were making up 62 percent of commercial banks' earnings, up from 33 percent in 1987.

As a result, what would have been a $300 billion sub-prime mortgage crisis when the bubble inevitably burst, turned into a multi-*trillion* dollar catastrophe engulfing the financial systems of the US and many other countries as well.

Between July 2004 and July 2006, the Fed began to pursue policies designed to raise interest rates on bank loans. This contributed to an increase in 1-year and 5-year adjustable mortgage rates, pushing up mortgage payments for many homeowners. Since asset prices generally move inversely to interest rates, it suddenly became riskier to speculate in housing. The bubble began deflating.

What Goes Up...

In early 2007 home foreclosure rates nosed upward and the US sub-prime mortgage industry simply collapsed, with more than 25 lenders declaring bankruptcy, announcing significant losses, or putting themselves up for sale.

The whole scheme had worked fine as long as the underlying collateral (homes) appreciated in value year after year. But as soon as house prices peaked, the upside-down pyramid of property, debt, CDOs, and derivatives wobbled and began crashing down.

For a brief time between 2006 and mid-2008 investors worldwide fled toward futures contracts in oil, metals, and food, driving up commodities prices. Food riots erupted in many poor nations, where the cost of wheat and rice doubled or tripled. In part, the boom was based on a fundamental economic trend: demand for commodities was growing—due in part to the expansion of economies in China, India, and Brazil—while supply growth was lagging. But speculation forced prices higher and faster than physical shortage could account for. For Western economies, soaring oil prices had a sharp recessionary impact, with already cash-strapped new homeowners now having to spend eighty to a hundred dollars every time

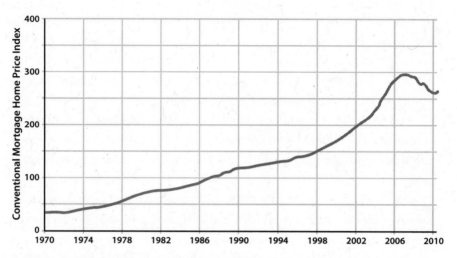

FIGURE 14. US Home Prices. Conventional Mortgage Home Price Index since 1970. Home prices rose consistently from 1970 until their peak in 2007, with the steepest rise occurring after 2000. Source: Freddie Mac.

they filled the tank in their SUV. The auto, airline, shipping, and trucking industries were suddenly reeling.

Between mid-2006 and September 2008, average US house prices declined by over 20 percent. As prices dove, many recent borrowers found themselves "underwater"—that is, with houses worth less than the amount of their loan; for those with adjustable-rate mortgages, this meant they could not qualify to refinance to avoid higher payments as interest rates on their loans reset. Default rates on home mortgages exploded. From 2006 to 2007, foreclosure proceedings increased 79 percent (affecting nearly 1.3 million properties). The trend worsened in 2008, with an 81 percent increase over the previous year and 2.3 million properties foreclosed. By August 2008, 9.2 percent of all US mortgages outstanding were either delinquent or in foreclosure; in September the following year, the figure had jumped to a whopping 14.4 percent.

Once property prices began to plummet and the subprime industry went bust, dominos throughout the financial world began toppling.

On September 15th, 2008, the entire financial system came within 48 hours of collapse. The giant investment house of Lehman Brothers went bankrupt, sending shock waves through global financial markets.[6] The

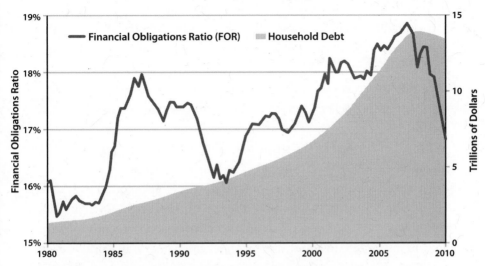

FIGURE 15. US Household Debt. Financial obligations ratio and total outstanding nominal debt of US households. A household's financial obligations ratio (FOR) is the ratio of its financial obligations (mortgage, consumer debt, automobile lease payments, rental payments on tenant-occupied property, homeowner's insurance, and property tax payments) to its disposable income. Just before the financial crisis, households were spending almost 19 percent of their disposable income on servicing their debt. Total outstanding household debt also peaked in 2008 just before the financial crisis at almost $14 trillion. To put this amount in perspective, the entire US economy was worth $14.3 trillion that same year. Source: The Federal Reserve.

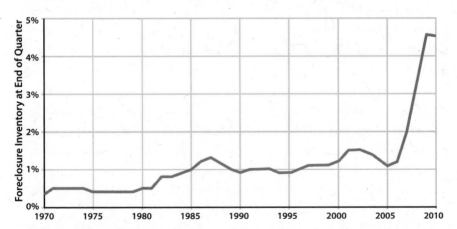

FIGURE 16. US Foreclosure Rate, 1970–2010. US foreclosure inventory at the end of each quarter. From 1970–2001, yearly averages are shown; from 2002–2010 quarterly data is shown. The foreclosure rate jumped dramatically during the financial crisis from 1.28 percent at the start of 2007 to 4.63 percent at the start of 2010, the highest level in the last forty years. Source: Mortgage Bankers Association, National Delinquency Survey, Foreclosure Inventory at End of Quarter.

global credit system froze, and the US government stepped in with an extraordinary set of bailout packages for the largest Wall Street banks and insurance companies. All told, the US package of loans and guarantees added up to an astounding $12 trillion. GDP growth for the nation as a whole went negative and eight million jobs disappeared in a matter of months.[7]

Much of the rest of the world was infected, too, due to interlocking investments based on mortgages. The Eurozone countries and the UK experienced economic contraction or dramatic slowing of growth; some developing countries that had been seeing rapid growth saw significant slowdowns (for example, Cambodia went from ten percent growth in 2007 to nearly zero in 2009); and by March 2009, the Arab world had lost an estimated $3 trillion due to the crisis — partly from a crash in oil prices.

Then in 2010, Greece faced a government debt crisis that threatened the economic integrity of the European Union. Successive Greek governments had run up large deficits to finance public sector jobs, pensions, and other social benefits; in early 2010, it was discovered that the nation's government had paid Goldman Sachs and other banks hundreds of millions of dollars in fees since 2001 to arrange transactions that hid the actual level of borrowing. Between January 2009 and May 2010, official government deficit estimates more than doubled, from 6 percent to 13.6 percent of GDP — the latter figure being one of the highest in the world. The direct effect of a Greek default would have been small for the other European economies, as Greece represents only 2.5 percent of the overall Eurozone economy — but it could have caused investors to lose faith in other European countries that also have high debt and deficit issues: Ireland, with a government deficit of 14.3 percent of GDP, the UK with 12.6 percent, Spain with 11.2 percent, and Portugal with 9.4 percent, were most at risk. And so Greece was bailed out with loans from the EU and the IMF, whose terms included the requirement to slash social spending.

By late November of 2010, it was clear that Ireland needed a bailout, too — and it got one, along with its own painful austerity package and loads of political upheaval. But this raised the inevitable questions: Who would be next? Could the IMF and the EU afford to bail out Spain if necessary? What would happen if the enormous UK economy needed rescue?

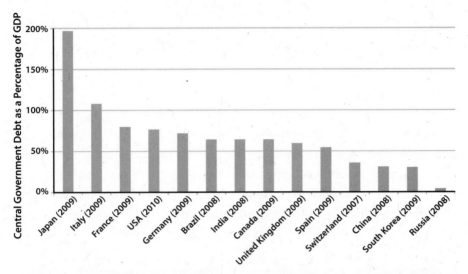

FIGURE 17A. Central Government Debt as a Percentage of GDP for Various Countries. High levels of government debt burden countries around the world, not just the US. For example, the debt of the Japanese government amounts to almost 200% of its GDP. Sources: McKinsey Global Institute, "Debt and deleveraging: The global credit bubble and its economic consequences," January 2010; The Federal Reserve.

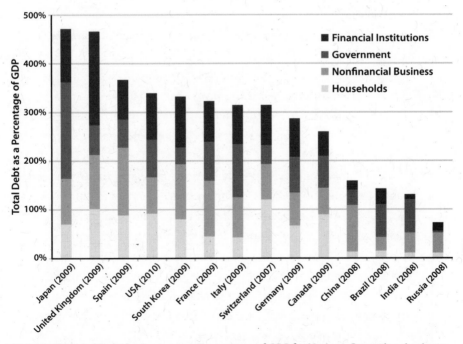

FIGURE 17B. Total Debt by Sector as a Percentage of GDP for Various Countries. Again we can see that the US is not alone when it comes to high levels of debt. The total debt of Japan and the UK amounts to around 450 percent of their respective GDP. Sources: McKinsey Global Institute, "Debt and deleveraging: The global credit bubble and its economic consequences," January 2010; The Federal Reserve.

Meanwhile China—whose economy continued growing at a scorching 10 percent per year, and which had run a large trade surplus for the past three decades—had inflated its own enormous real estate bubble. Average housing prices in the country tripled from 2005 to 2009; and price-to-income and price-to-rent ratios for property, as well as the number of unoccupied residential and commercial units, were all sky-high.

In short, a global economy that had appeared robust and stable in 2007 was suddenly revealed to be very fragile, suffering from several persistent maladies—any one of which could erupt into virulence, spreading rapidly and sending the world back into the throes of crisis.

BOX 2.1 | Plenty of Blame to Go Around

The bipartisan Financial Crisis Inquiry Commission (established by Congress as part of the Fraud Enforcement and Recovery Act of 2009) released its report in January 2011. The many causal factors it highlighted include:

- Federal Reserve Chairman (1987–2006) Alan Greenspan's refusal to perform his regulatory duties because he did not believe in them. Greenspan allowed the credit bubble to expand, driving housing prices to dangerously unsustainable levels while advocating financial deregulation. The Commission called this a "pivotal failure to stem the flow of toxic mortgages" and "the prime example" of government negligence.
- Federal Reserve Chairman (2006-present) Ben Bernanke's failure to foresee the crisis.
- The Bush administration's "inconsistent response" in saving one financial giant—Bear Stearns—while allowing another—Lehman Brothers—to fail; this "added to the uncertainty and panic in the financial markets."
- Bush Treasury Secretary Henry Paulson Jr.'s failure to understand the magnitude of the problem with subprime mortgages.
- The Clinton White House's (and Treasury Secretary Lawrence Summers's) crucial error in shielding over-the-counter derivatives from regulation in the Commodity Futures Modernization Act; this constituted "a key turning point in the march toward the financial crisis."

- Then NY Fed President, now Treasury Secretary Timothy F. Geithner's failure to "clamp down on excesses by Citigroup in the lead-up to the crisis."
- The Fed's maintenance of low interest rates long after the 2001 recession, which "created increased risks."
- The financial sector's spending of $2.7 billion on lobbying from 1999 to 2008, with members of Congress affiliated with the industry raking in more than $1 billion in campaign contributions.
- The credit-rating agencies' stamping of "their seal of approval" on securities that proved to be far more risky than advertised (because they were backed by mortgages provided to borrowers who were unable to make payments on their loans).
- The Securities and Exchange Commission's permitting of the five biggest banks to ramp up their leverage, hold insufficient capital, and engage in risky practices.
- The nation's five largest investment banks' buildup of wildly excessive leverage: They kept only $1 in capital to cover losses for about every $40 in assets.
- The Office of the Comptroller of the Currency's (along with the Office of Thrift Supervision's) blocking of state regulators from reining in lending abuses.
- "Questionable practices by mortgage lenders and careless betting by banks."
- The "bumbling incompetence among corporate chieftains" as to the risk and operations of their own firms. Among corporate heads at the large financial firms (including Citigroup, AIG, and Merrill Lynch), the panel says its examination found "stunning instances of governance breakdowns and irresponsibility."

Commission members disagreed on the significance of the roles of Freddie Mac and Fannie Mae in the crisis.

The Commission has indicated that it will make criminal referrals.[8]

The Mother of All Manias

The US real estate bubble of the early 2000s was the largest in history (in terms of the amount of capital involved).[9] And its crash carried an eerie echo of the 1930s: some economists have argued that it wasn't just the stock market crash that drove the Great Depression, but also cascading farm failures, which made it impossible for farmers to make mortgage payments — along with housing bubbles in Florida, New York, and Chicago.[10]

Real estate bubbles are essentially credit bubbles, because property owners generally use borrowed money to purchase property (this is in contrast to currency bubbles, in which nations inflate their currency to pay off government debt). The amount of outstanding debt soars as buyers flood the market, bidding property prices up to unrealistic levels and taking out loans they cannot repay. Too many houses and offices are built, and materials and labor are wasted in building them. Real estate bubbles also lead to an excess of homebuilders, who must retrain and retool when the bubble bursts. These kinds of bubbles lead to systemic crises affecting the economic integrity of nations.[11]

Indeed, the housing bubble of the early 2000s had become the oxygen of the US economy — the source of jobs, the foundation for Wall Street's recovery from the dot-com bust, the bait for foreign capital, and the basis for household wealth accumulation and spending. Its bursting changed everything.

And there is reason to think it has not yet fully deflated: commercial real estate may be waiting to exhale next. Over the next five years, about $1.4 trillion in US commercial real estate loans will reach the end of their terms and require new financing. Commercial property values have fallen more than 40 percent nationally since their 2007 peak, so nearly half the loans are underwater. Vacancy rates are up and rents are down.

The impact of the real estate crisis on banks is profound, and goes far beyond defaults upon outstanding mortgage contracts: systemic dependence on MBSs, CDOs, and derivatives means many of the banks, including the largest, are effectively insolvent and unable to take on more risk (we'll see why in more detail in the next section).

Demographics do not favor a recovery of the housing market anytime soon. The oldest of the Baby Boomers are 65 and entering retirement.

Few have substantial savings; many had hoped to fund their golden years with house equity — and to realize that, they must sell. This will add more houses to an already glutted market, driving prices down even further.

In short, real estate was the main source of growth in the US during the past decade. With the bubble gone, leaving a gaping hole in the economy, where will new jobs and further growth come from? Can the problem be solved with yet another bubble?

BOX 2.2   How to Create a Financial Crisis

In their IMF Working Paper, "Inequality, Leverage and Crises," Michael Kumhof and Romain Rancière construct a simple model for financial crises with the following narrative: (a) growing inequality produces less money for the middle class and more money for the wealthy; (b) the rich loan much of this money back to the middle class so they can continue to improve their living standards even with stagnant incomes; (c) the financial sector expands to mediate all this; and (d) this eventually results in a credit crisis. Kumhof and Rancière write, in summary:

"This paper has presented stylized facts and a theoretical framework that explore the nexus between increases in the income advantage enjoyed by high income households, higher debt leverage among poor and middle income households, and vulnerability to financial crises. This nexus was prominent prior to both the Great Depression and the recent crisis. In our model it arises as a result of increases in the bargaining power of high income households. The key mechanism, reflected in a rapid growth in the size of the financial sector, is the recycling of part of the additional income gained by high income households back to the rest of the population by way of loans, thereby allowing the latter to sustain consumption levels, at least for a while. But without the prospect of a recovery in the incomes of poor and middle income households over a reasonable time horizon, the inevitable result is that loans keep growing, and therefore so does leverage and the probability of a major crisis that, in the real world, typically also has severe implications for the real economy."[12]

This dynamic is also occurring between rich nations and poor nations.

Limits to Debt

Let's step back a moment and look at our situation from a slightly different angle. Take a careful look at Figure 18, the total amount of debt extant each year in the US since 1979. The graph breaks the debt down into four categories — household, corporate, financial sector, and government. All have grown very substantially during these past 30+ years, with the largest percentage growth having taken place in the financial sector. Note the shape of the curve: it is not a straight line (which would indicate additive growth); instead, up until 2008, it more closely resembles the J-curve of compounded or exponential growth (as discussed in the Introduction).

Growth that proceeds this way, whether it's growth in US oil production from 1900 to 1970 or growth in the population of *Entamoeba histolytica* in the bloodstream of a patient with amoebic dysentery, always hits hard limits eventually.

With regard to debt, what are those limits likely to be and how close are we to hitting them?

A good place to start the search for an answer would be with an ex-

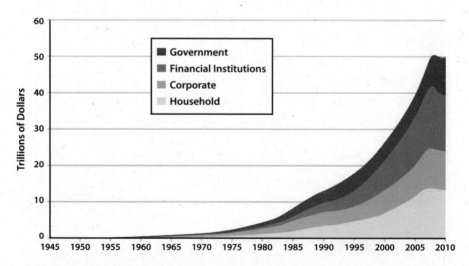

FIGURE 18. Total US Debt, 1945–2010. US debt by sector in nominal values (not inflation adjusted). We see the rapid expansion of both household and financial sector debt beginning in 2000, spurred by low interest rates and rising home values. Starting in 2008, household and financial debt contract, while government debt expands. Source: The Federal Reserve, Z.1 Flow of Funds Accounts of the United States.

ploration of how we have managed to grow our debt so far. It turns out that, in an economy that's based on money creation through fractional reserve banking, with ever more loans being taken out to finance ever more consumer purchases and capital projects, it is usually possible to repay earlier debts along with the interest attached to those debts. There is never enough money in the system at any one time to repay *all* outstanding debt with interest; but, as long as total debt (and therefore the money supply as well) is constantly growing, that doesn't pose a practical problem. The system as a whole does have some of the characteristics of a bubble or a Ponzi scheme, but it also has a certain internal logic and even the potential for (temporary) dynamic stability.

However, there are practical limits to debt within such a system, and those limits are likely to show up in somewhat different ways for each of the four categories of debt indicated in the graph.

Government Debt

With government debt, problems arise when required interest payments become a substantial fraction of tax revenues. Let's start with some basics:

- *Government debt* is the total of what the government owes.
- Payment of government debt can obviously be delayed, but controversy exists over how long payment can reasonably be delayed.
- Government debt results, of course, in *interest payments*. Every year federal *revenues* must be used to pay interest on the government debt (which was incurred in the past). There are usually disagreements on whether the interest was incurred for a good purpose, but everyone agrees that it is important to know exactly how much money is owed in interest when planning for the future.
- Both government debt and interest payments can increase. If the government spends more than it takes in during a specific year, a shortfall develops. That shortfall is referred to as the *deficit*. The government handles the deficit by borrowing more money (at interest). Thus the deficit is the shortfall for a specific year, and the government debt is the total of those shortfalls.
- A deficit not only adds to *government debt* (that IOU over there in the corner) but it adds to the *interest* payments (right here, not over there

in the corner) that must be made every year. Those interest payments are made with the *tax revenues* that the government collects every year, or with more borrowing.

Currently for the US, the total Federal budget amounts to about $3.5 trillion, of which 12 percent (or $414 billion) goes toward interest payments. But in 2009, tax revenues amounted to only $2.1 trillion; thus interest payments currently consume almost 20 percent, or nearly one-fifth, of tax revenues. For various reasons (including the economic recession, the wars in Iraq and Afghanistan, the Bush tax cuts, and various stimulus programs) the Federal government is running a deficit of over a trillion dollars a year currently. That adds to the debt, and therefore to future interest payments. Government debt stands at over $14 trillion now (it has increased by more than 50 percent since 2006).[13] By the time the debt reaches $20 trillion, probably only a few years from now, interest payments may constitute the largest Federal budget outlay category, eclipsing even military expenditures.[14] If Federal revenues haven't increased by that time, government debt interest payments will be consuming 20 percent of them. Interest already eats up nearly half the government's income tax receipts, which are estimated at $901 billion for fiscal year 2010.[15]

Clearly, once 100 percent of government revenues have to go toward interest payments and all government operations have to be funded with more borrowing — on which still more interest will have to be paid — the system will have arrived at a kind of financial singularity: a black hole of debt, if you will. But in all likelihood we would not have to get to that ultimate impasse before serious problems appear. Many economic commentators suggest that when government has to spend 30 percent of tax receipts on interest payments, the country is in a debt trap from which there is no easy escape. Given current trajectories of government borrowing and interest rates, that 30 percent mark could be hit in just a few years. Even before then, US credit worthiness will take a beating.

However, some argue that limits to government debt (due to snowballing interest payments) need not be a hard constraint — especially for a large nation, like the US, that controls its own currency.[16] The United States government is constitutionally empowered to create money, including creating money to pay the interest on its debts. Or, the government could in effect loan the money to itself via its central bank, which would then rebate interest payments back to the Treasury (this is in fact what the Treasury and Fed are doing with Quantitative Easing 2, discussed below).[17]

The most obvious complication that might arise is this: If at some point general confidence that external US government debt (i.e., money owed to private borrowers or other nations) will be repaid with debt of equal "value" were deeply and widely shaken, potential buyers of that debt might decide to keep their money under the metaphorical mattress (using it to buy factories or oilfields instead), even if doing so posed its own set of problems. Then the Fed would become virtually the only available buyer of government debt, which might undermine confidence in the US dollar, possibly igniting a rapid spiral of refusal that would end only when the currency failed. There are plenty of historic examples of currency failures, so this would not be a unique occurrence.[18]

Some who come to understand that government deficit spending is unsustainable immediately conclude that the sky is falling and doom is imminent. It is disquieting, after all, to realize for the first time that the world economic system is a kind of Ponzi scheme that is only kept going by the confidence of its participants. But as long as deficit spending doesn't exceed certain bounds, and as long as the economy resumes growth in the not-too-distant future, then the scheme can be sustained for quite some time. In fact, Ponzi schemes theoretically can continue forever — if the number of potential participants is infinite. The absolute size of government debt is not necessarily a critical factor, as long as future growth will be sufficient so that the proportion of debt relative to revenues remains the same. Even an increase in that proportion is not necessarily cause for alarm, as long as it is only temporary. This, at any rate, is the Keynesian argument. Keynesians would also point out that government debt is only one category of total debt, and that US government debt hasn't grown proportionally relative to other categories of debt to any alarming degree (until the current recession). Again, as long as growth returns, further borrowing can be justified (up to a point) — especially if the goal is to restart growth.[19]

The risks of increasing government debt can be summarized as: (a) rising interest costs, (b) loss of credit-worthiness, and (c) potential currency failure.

Household Debt

The limits to household debt are different, but somewhat analogous: consumers can't create money the way banks (and some governments) do, and can't take on more debt if no one will lend to them. Lenders usually require collateral, so higher net worth (often in the form of house equity) translates to greater ability to take on debt; likewise, lenders wish to see evidence of ability to make payments, so a higher salary also translates to a greater ability to take on increased levels of debt.

As we have seen, the actual inflation-adjusted income of American workers has not risen substantially since the 1970s, but home values did rise during the 2000–2006 period, giving many households a higher theoretical net worth. Many homeowners used their soaring house value as collateral for more debt — in many cases, substantially more. At the same time, lenders found ways of easing consumer credit standards and making credit generally more accessible — whether through "no-doc" mortgages or blizzards of credit card offers. The result: household debt increased from less than $2 trillion in 1980 to $13.5 trillion in 2008. This borrowing and spending on the part of US households was not only the major engine of domestic economic expansion during most of the last decade, but a major component of worldwide economic growth as well.

But with the crash in the US real estate market starting in 2007, household net worth also crashed (falling by a total of $17.5 trillion or 25.5 percent from 2007 to 2009 — equivalent to the loss of one year of GDP); and as unemployment rose from 4.6 percent in 2007 to almost ten percent (as officially measured) in 2010, average household income declined. At the same time, banks tightened their lending standards, with credit card companies slashing the number of offers and mortgage lenders requiring much higher qualifications from borrowers. Thus the ability of households to take on more debt has contracted substantially. Less debt means less spending (households usually borrow money so they can spend it — whether for a new car or a kitchen makeover). This is potentially a short-term problem; however, the only way the situation will change is if somehow the economy as a whole begins to grow again (leading to higher house prices, lower unemployment, and easier credit). Here's the catch:

increased consumer demand is a big part of what would be needed to drive that shift back to growth.

So we just need to get households borrowing and spending again. Perhaps government could somehow put a bit of seed money in citizens' pockets ("Cash for Clunkers," anyone?) to start the process. Even if that doesn't work, at some point consumers will have paid down (or defaulted on) their debts sufficiently so that they will want to borrow more. But, again, demographics suggest this would be a long wait: as mentioned earlier, Baby Boomers (the most numerous demographic cohort in the nation's history, encompassing 70 million Americans) are reaching retirement age, which means that their lifetime spending cycle has peaked. It's not that Boomers won't continue to buy things (everybody has to eat), but their aggregate spending is unlikely to increase, given that cohort members' savings are, on average, inadequate for retirement (one-third of them have no savings whatever). Out of necessity, Boomers will be saving more from now on, and spending less. And that won't help the economy grow. We may not have hit a hard, final, and axiomatic limit to household debt, but (in the US, at least) there is no realistic basis for a resumption of rates of growth in borrowing and spending seen in recent decades.

Corporate Debt

When demand for products declines, corporations aren't inclined to borrow to increase their productive capacity. Even corporate borrowing aimed at increasing financial leverage has limits. Too much corporate debt reduces resiliency during slow periods — and the future is looking slow for as far as the eye can see. Durable goods orders are down, housing starts and new home sales are down, savings are up. As a result, banks don't *want* to lend to companies, because the risk of default on such loans is now perceived as being higher than it was a few years ago; in addition, the banks are reluctant to take on more risk of any sort given the fact that many of the assets on their balance sheets consist of now-worthless derivatives and CDOs. Nevertheless, corporate debt levels hit all-time highs in 2010.

Meanwhile, ironically and perhaps surprisingly, US corporations are sitting on over a trillion dollars of ready cash because they cannot identify profitable investment opportunities and because they want to hang onto whatever cash they have in anticipation of continued hard times.

If only we could get to the next upside business cycle, then more corporate debt would be justified for both lenders and corporate borrowers. But so far confidence in the future is still weak.

Financial Sector Debt

The category of financial sector debt — which, of the four categories, has grown the most — consists of debt and leverage within the financial system itself. This category can in principle be disregarded, as financial institutions are primarily acting as intermediaries for ultimate borrowers. However, in this case, standing on principle does not aid comprehension. We are not including within this category the notional value of derivatives contracts, which is roughly five times the amount of US government, household, corporate, and financial debt combined (roughly $260 trillion in outstanding derivates, versus $55 trillion in debt). But while this category does not directly include the value of derivatives, the expansion of the financial sector has largely been based on derivatives trading. And derivatives have arguably helped create a situation that limits further growth in the financial system's ability to perform its only truly useful function within society — to provide investment capital for productive enterprise.

One of the main reforms enacted during the Great Depression, contained in the Glass Steagall Act of 1933, was a requirement that commercial banks refrain from acting as investment banks. In other words, they were prohibited from dealing in stocks, bonds, and derivatives. This prohibition was based on an implicit understanding that there should be some sort of firewall within the financial system separating productive investment from pure speculation, or gambling. This firewall was eliminated by the passage of the Gramm–Leach–Bliley Act of 1999 (for which the financial services industry lobbied tirelessly). As a result, all large US banks have for the past decade become deeply engaged in speculative investment, using both their own and their clients' money.

With derivatives, since there is no requirement to own the underlying asset, other than a small percentage of its notional value, and since there is often no requirement of evidence of ability to cover the bet, there is no effective limit to the amount that can be wagered. It's true that many derivatives largely cancel each other out, and that their ostensible purpose is to reduce financial risk. Nevertheless, if a contract is settled, somebody has to pay — unless they can't.

Credit default swaps (CDSs, discussed in the last chapter) are usually traded "over the counter" — meaning without the knowledge of anyone other than the two counterparties; they are a sort of default insurance: a contract holder acts as "insurer" against default, bankruptcy, or other "credit event," and collects regular "insurance" payments as premiums; this comes as "free money" to the "insurer." But if default occurs, then a huge payment becomes due. Perversely, it is perfectly acceptable to take out a credit default swap on someone else's debt. Here's one example: In 2005, auto parts maker Delphi defaulted on $5.2 billion in outstanding bonds and loans — but over $20 billion in credit default derivative contracts had been written on those bonds and loans. The result: massive losses on the part of derivative holders, much more than for those who held the bonds or loans. This degree of leverage was not uncommon throughout corporate America, and the US financial system as a whole. Were derivatives really reducing risk, or merely spreading it throughout the economy?

An even more telling example relates to the insurance giant AIG, which insured the obligations of various financial institutions through CDSs. The transaction went like this: AIG received a periodic premium in exchange for a promise to pay party A if party B defaulted. As it turned out, AIG did not have the capital to back its CDS commitments when defaults began to spread throughout the US financial system in 2008, and a failure of AIG would have brought down many other companies in a kind of financial death-spiral. Therefore the Federal government stepped in to bail out AIG with tens of billions of dollars.

In the heady years of the 2000s, even the largest and most prestigious banks engaged in what can only be termed criminal behavior on a massive scale. As revealed in sworn Congressional testimony, firms including Goldman Sachs deliberately created flawed securities and sold tens of billions of dollars' worth of them to investors, then took out many more billions of dollars' worth of derivatives contracts essentially betting against the securities they themselves had designed and sold. They were quite

simply defrauding their customers, which included foreign and domestic
pension funds. To date, no senior executive with any bank or financial
services firm has been prosecuted for running these scams. Instead, most
of the key figures are continuing to amass immense personal fortunes,
confident no doubt that what they were doing—and in many cases continue to do—is merely a natural extension of the inherent logic of their
industry.

The degree and concentration of exposure on the part of the biggest banks with regard to derivatives was and is remarkable: As of 2005,
JP Morgan Chase, Bank of America, Citibank, Wachovia, and HSBC together accounted for 96 percent of the $100 *trillion* of derivatives contracts
held by 836 US banks.[20]

Even though many derivatives were insurance against default, or
wagers that a particular company would fail, to a large degree they constituted a giant hedged bet that the economy as a whole would continue to
grow (and, more specifically, that the value of real estate would continue to
climb). So when the economy *stopped* growing, and the real estate bubble
began to deflate, this triggered a systemic unraveling that could be halted
(and only temporarily) by massive government intervention.

Suddenly "assets" in the form of derivative contracts that had a stated
value on banks' ledgers were clearly worth much less. If these assets had
to be sold, or if they were "marked to market" (valued on the books at the
amount they could actually sell for), the banks would be shown to be insolvent. Government bailouts essentially enabled the banks to keep those
assets hidden, so that banks could appear solvent and continue carrying
on business.

Despite the proliferation of derivatives, the financial system still largely
revolves around the timeworn practice of receiving deposits and making
loans. Bank loans are the source of money in our modern economy. If the
banks go away, so does the rest of the economy (at least temporarily, until
the functions of the banks can be taken up by other institutions).

But as we have just seen, many banks are probably actually insolvent
because of the many near-worthless derivative contracts and bad mortgage loans they count as assets on their balance sheets.

One might well ask: *If commercial banks have the power to create money, why can't they just write off these bad assets and carry on*? Ellen Brown explains the point succinctly in her useful book *The Web of Debt*:

> [U]nder the accountancy rules of commercial banks, all banks are obliged to balance their books, making their assets equal their liabilities. They can *create* all the money they can find borrowers for, but if the money isn't paid back, the banks have to record a loss; and when they cancel or write off debt, their assets fall. To balance their books...they have to take the money either from profits or from funds invested by the bank's owners [i.e., shareholders]; and if the loss is more than its owners can profitably sustain, the bank will have to close its doors.[21]

So, given their exposure via derivatives, bad real estate loans, and MBSs, the banks aren't making new loans because they can't take on more risk. The only way to reduce that risk is for government to guarantee the loans. Again, as long as the down-side of this business cycle is short, such a plan could work in principle.

But whether it actually will work in the current situation is problematic. As noted above, Ponzi schemes can theoretically go on forever, as long as the number of new investors is infinite. Yet in the real world the number of potential investors is always finite. There are limits. And when those limits are hit, Ponzi schemes can unravel very quickly.

All Loaned Up and Nowhere to Go

These are the four categories of debt. Over the short term, there is no room for growth of debt in the household or corporate sectors. Within the financial sector, there is little room for growth in productive lending. The shadow banks can still write more derivative contracts, but that doesn't do anything to help the real economy and just spreads risk throughout the system. That leaves government, which (if it controls its own currency and can fend off attacks from speculators) can continue to run large deficits, and the central banks, which can enable those deficits by purchasing government debt outright. But unless such efforts succeed in jump-starting growth in the other sectors, this is just a temporary end-game strategy.

A single statistic is revealing: in the US, the ratio of total debt to GDP has risen to more than 300 percent, exceeding the previous record of 290 percent achieved immediately prior to the stock market crash of 1929.[22] If there is a theoretical or practical limit to debt, the US seems destined to reach it, and soon.

Remember: in a system in which money is created through bank loans, there is never enough money in existence to pay back all debts with interest. The system only continues to function as long as it is growing.[23]

So, what happens to the existing mountain debt in the absence of economic growth? Answer: Some kind of debt crisis. And that is what we are seeing.

Debt crises have occurred throughout the history of civilizations, beginning long before the invention of fractional reserve banking and credit cards. Many societies learned to solve the problem with a "debt jubilee": According to the Book of Leviticus in the Bible, every fiftieth year is a Jubilee Year, in which slaves and prisoners are to be freed and debts are to be forgiven. Evidence of similar traditions can be found in an ancient Hittite-Hurrian text entitled "The Song of Debt Release"; in the history of Ancient Athens, where Solon (638–558 BCE) instituted a set of laws called *seisachtheia*, canceling all current debts and retroactively canceling previous ones that had caused slavery and serfdom (thus freeing debt slaves and debt serfs); and in the Qur'an, which advises debt forgiveness for those who are genuinely unable to pay.

For householders facing unaffordable mortgage payments or a punishing level of credit card debt, a jubilee may sound like a splendid idea. But what would that actually mean today, if carried out on a massive scale — when debt has become the very fabric of the economy? Remember: we have created an economic machine that needs debt like a car needs gas.

Realistically, we are unlikely to see a general debt jubilee in coming years (though we'll reconsider that possibility in more detail in Chapter 6); what we may see instead are defaults and bankruptcies that accomplish essentially the same thing — the destruction of debt. Which, in an economy like ours, effectively means a destruction of wealth and claims upon wealth. Debt would have to be written off in enormous amounts — by the trillions of dollars. Over the short term, government could attempt to stanch this flood of debt-shedding in the household, corporate, and financial sectors by taking on more debt of its own—but eventually it might not be able to keep up, given the inherent limits on government borrowing discussed above. Central banks could also help keep banks' toxic assets hidden, a strategy the Fed seems in fact to be pursuing, though it is one not likely to succeed indefinitely.

We began with the question, "How close are we to hitting the limits to debt?" The evident answer is: we have already probably hit realistic limits to household debt and corporate debt; the ratio of US total debt-to-GDP is probably near or past the danger mark; and limits to government debt may be within sight, though that conclusion is more controversial.

Stimulus Duds, Bailout Blanks

In response to the financial crisis, governments and central banks have undertaken a series of extraordinary, dramatic measures. In this section we will focus primarily on the US (the bailouts of banks, insurance and car companies, and Government Sponsored Enterprises—i.e., Fannie Mae and Freddie Mac; the stimulus packages of 2008 and 2009; and actions by, and new powers given to the Federal Reserve); later we will also briefly touch upon some actions by governments and central banks in other nations (principally China and the Eurozone).

For the US, actions undertaken by the Federal government and the Federal Reserve bank system have so far resulted in totals of $3 trillion actually spent and $11 trillion committed as guarantees. Some of these actions are discussed below; for a complete tally of the expenditures and commitments, see the online CNN Bailout Tracker.[24]

Bailouts

Bailouts directly funded by the US Department of the Treasury were mostly bundled together under the Troubled Assets Relief Program (TARP), signed into law October 3, 2008, which allowed the Treasury to purchase or insure up to $700 billion worth of "troubled assets." These were defined as residential or commercial mortgages and "any securities, obligations, or other instruments that are based on or related to such mortgages," issued on or before March 14, 2008. Essentially, TARP allowed the Federal government to purchase illiquid, difficult-to-value assets (primarily CDOs) from banks and other financial institutions in order to prevent a wave of insolvency from sweeping the financial world. The list of companies receiving TARP funds included the largest, wealthiest, and most powerful firms on Wall Street—Citigroup, Bank of America, AIG, JPMorgan Chase, Wells Fargo, Goldman Sachs, and Morgan Stanley—as well as GMAC, General Motors, and Chrysler.

The program was controversial, with some calling it "lemon socialism" (privatization of profits and socialization of losses). Critics were especially outraged when it became known that executives in the bailed-out companies were continuing to reward themselves with enormous salaries and bonuses. Some instances of fraud were uncovered, as well as the use of substantial amounts of money by participating companies to lobby against financial reforms.

Nevertheless, some of the initial fears about good money being thrown after bad did not appear to be borne out. Much of the TARP outlay was quickly repaid (for example, as of mid-2010, over $169 billion of the $245 billion invested in US banks had been paid back, including $13.7 billion in dividends, interest and other income). Some of the repayment efforts appeared to be motivated by the desire on the part of companies to get out from under onerous restrictions (including restrictions by the Obama administration on executive pay).

A bailout of Fannie Mae and Freddie Mac was announced in September 2008 in which the federal government, via the Federal Housing Finance Agency, placed the two firms into conservatorship, dismissed the firms' chief executive officers and boards of directors, and made the Treasury 79.9 percent owners of each GSE. The authority of the US Treasury to continue paying to stabilize Fannie Mae and Freddie Mac is limited only by statutory constraints to Federal government debt. The Fannie-Freddie bailout law increased the national debt ceiling $800 billion, to a total of $10.7 trillion, in anticipation of the potential need for government mortgage purchases.

The US market for mortgage-backed securities had collapsed from $1.9 trillion in 2006 to just $50 billion in 2008. Thus the upshot of the Freddie-

Fannie bailout was that the Federal government became the US mortgage lender of first and last resort.

Altogether, the bailouts succeeded in preventing an immediate meltdown of the national (and potentially the global) financial system. But they did not significantly alter the culture of Wall Street (i.e., the paying of exorbitant bonuses for the acquisition of inappropriate risk via cutthroat competition that ignores long-term sustainability of companies or economies). And they did not relieve the underlying solvency crisis faced by the banks — they merely papered these problems over temporarily, until the remaining bulk of the "troubled" assets are eventually marked to market (listed on banks' balance sheets at realistic values). Meanwhile, the US government has taken on the burden of guaranteeing most of the nation's mortgages, in a market in which residential and commercial real estate values may be set to decline further.

Stimulus Packages

During 2008 and 2009, the US Federal government implemented two stimulus packages, spending a total of nearly $1 trillion.

The first (the Economic Stimulus Act of 2008) consisted of direct tax rebates, mostly distributed at $300 per taxpayer, or $600 per couple filing jointly. The total cost of the bill was projected at $152 billion.

The second, the American Recovery and Reinvestment Act of 2009, or ARRA, was comprised of an enormous array of projects, tax breaks, and programs — everything from $100 million for free school lunch programs to $6 billion for the cleanup of radioactive waste, mostly at nuclear weapons production sites. The total nominal worth of the spending package was $787 billion. A partial list:

- Tax incentives for individuals (e.g., a new payroll tax credit of $400 per worker and $800 per couple in 2009 and 2010). Total: $237 billion.
- Tax incentives for companies (e.g., to extend tax credits for renewable energy production). Total: $51 billion.
- Healthcare (e.g., Medicaid). Total: $155.1 billion.
- Education (primarily, aid to local school districts to prevent layoffs and cutbacks). Total: $100 billion.

- Aid to low-income workers, unemployed, and retirees (including job training). Total: $82.2 billion ($40 billion of this went to provide extended unemployment benefits through Dec. 31, and to increase them).
- Infrastructure Investment. Total: $105.3 billion.
- Transportation. Total: $48.1 billion.
- Water, sewage, environment, and public lands. Total: $18 billion.

In addition to these two programs, Congress also appropriated a total of $3 billion for the temporary Car Allowance Rebate System (CARS) program, known colloquially as "Cash for Clunkers," which provided cash incentives to US residents to trade in their older gas guzzlers for new, more fuel-efficient vehicles.

The New Deal had cost somewhere between $450 and $500 billion and had increased government's share of the national economy from 4 percent to 10 percent. ARRA represented a much larger outlay that was spent over a much shorter period, and increased government's share of the economy from 20 percent to 25 percent.

Given the scope and cost of the two stimulus programs, they were bound to have some effect — though the extent of the effect was debated mostly along political lines. The 2008 stimulus helped increase consumer spending (one study estimated that the stimulus checks increased spending by 3.5 percent).[25] And unemployment undoubtedly rose less in 2009–2010 than it would have done without ARRA.

Whatever the degree of impact of these spending programs, it appeared to be temporary. For example, while "Cash for Clunkers" helped sell almost 700,000 cars and nudged GM and Chrysler out of bankruptcy, once the program expired US car sales languished at their lowest level in 30 years.

At the end of 2010, President Obama and congressional leaders negotiated a compromise package of extended and new tax cuts that, in total, would reduce potential government revenues by an estimated $858 billion. This was, in effect, a third stimulus package.

Critics of the stimulus packages argued that transitory benefits to the economy had been purchased by raising government debt to frightening levels.[26] Proponents of the packages answered that, had government not

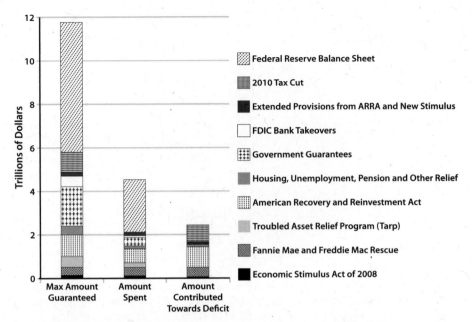

FIGURE 19A. Stimulus and Bailouts of 2008–2010. Since 2008, the Federal Government has allocated close to $12 trillion for stimulus and bailout programs. However, all this money has not actually been spent. So far, $4.6 trillion has been spent to stabilize the economy. The contribution of stimulus and bailout spending towards the deficit is still smaller, just over $2 trillion; the reason being, in part, that the actions of the Federal Reserve (bank guarantees, loans, and asset purchases) are not considered a contribution towards the deficit. Also, most of the TARP money has since been repaid. Source: The Committee for a Responsible Federal Budget.

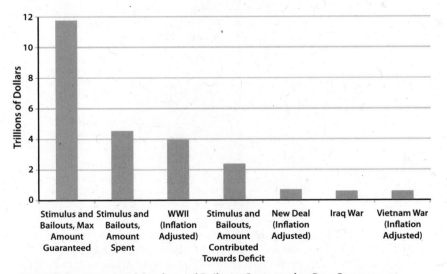

FIGURE 19B. 2008–2010 Stimulus and Bailouts Compared to Past Government Spending. The stimulus and bailouts of 2008–2010 dwarf most previous federal expenditures for a single purpose, exceeding even US spending for WWII. Source: The Committee for a Responsible Federal Budget, *Los Angeles Times*, Forbes.com.

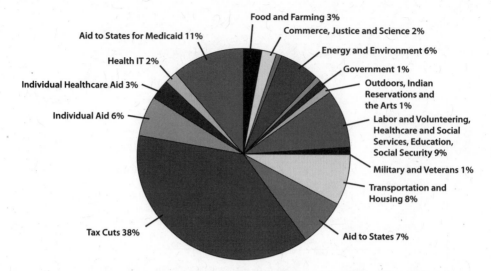

FIGURE 20A. American Recovery and Reinvestment Act of 2009. Allocation of funds of the American Recovery and Reinvestment Act of 2009. The economic stimulus bill passed by the Obama administration totaled around $790 billion, of which $665 has been spent so far. Of the $790 billion, close to 40 percent came not in the form of government spending, but rather in the form of lost revenues as a result of tax cuts. The second largest expenditure of the stimulus was the $90 billion allocated to states for Medicaid programs. Source: *The Wall Street Journal*, "Getting to $787 Billion," February 17, 2009.

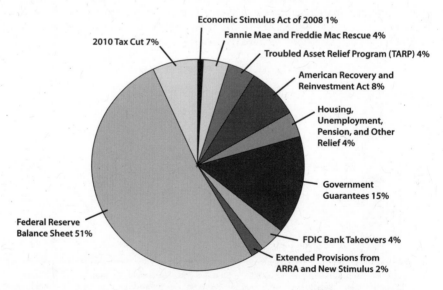

FIGURE 20B. Stimulus and Bailouts — Maximum Amount Guaranteed. Of the $11.8 trillion in funds allocated by the federal government for stimulus, bailouts, and bank guarantees, almost three quarters of the money comes in the form of an expanded Federal Reserve Balance Sheet and government guarantees for banks and other financial institutions. As of February 2011, not all of this money has been spent — only $4.6 trillion, of which $2.5 trillion has been added to the deficit. The remainder is on tap, to be used at the discretion of the Federal Government and the Federal Reserve. Source: The Committee for a Responsible Federal Budget.

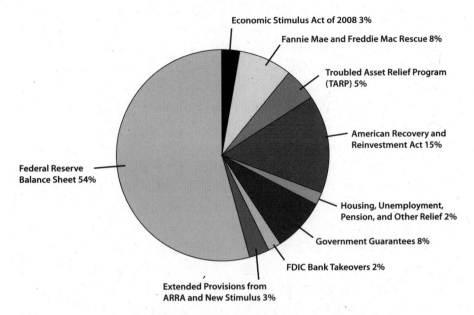

FIGURE 20C. Stimulus and Bailouts—Amount Spent. This chart shows a breakdown of the $4.6 trillion spent so far by the Federal Government to rescue the economy from collapse. The Federal Reserve spent the majority of the funds in order to stabilize systemically critical institutions. These expenditures took the form of loans, asset purchases, and guarantees. Source: The Committee for a Responsible Federal Budget.

acted so boldly, an economic crisis might have turned into complete and utter ruin.

Actions By, and New Powers of, the Federal Reserve

While the US government stimulus packages were enormous in scale, the actions of the Federal Reserve dwarfed them in terms of dollar amounts committed.

During the past three years, the Fed's balance sheet has swollen to more than $2 trillion through its buying of bank and government debt. Actual expenditures included $29 billion for the Bear Stearns bailout; $149.7 billion to buy debt from Fannie Mae and Freddie Mac; $775.6 billion to buy mortgage-backed securities, also from Fannie and Freddie; and $109.5 billion to buy hard-to-sell assets (including MBSs) from banks. However, the Fed committed itself to trillions more in insuring banks against losses, loaning to money market funds, and loaning to banks to purchase commercial paper. Altogether, these outlays and commitments totaled a minimum of $6.4 trillion.

Documents released by the Fed on December 1, 2010 showed that more than $9 trillion in total had been supplied to Wall Street firms, commercial banks, foreign banks, and corporations, with Citigroup, Morgan Stanley, and Merrill Lynch borrowing sums that cumulatively totaled over $6 trillion. The collateral for these loans was undisclosed but widely thought to be stocks, CDSs, CDOs, and other securities of dubious value.[27]

In one of its most significant and controversial programs, known as "quantitative easing," the Fed twice expanded its balance sheet substantially, first by buying mortgage-backed securities from banks, then by purchasing outstanding Federal government debt (bonds and Treasury certificates) to support the Treasury debt market and help keep interest rates down on consumer loans. The Fed essentially created money on the spot for this purpose (though no money was literally "printed").

In addition, the Federal Reserve has created new sub-entities to pursue various new functions:

- *Term Auction Facility* (which injects cash into the banking system),
- *Term Securities Lending Facility* (which injects Treasury securities into the banking system),
- *Primary Dealer Credit Facility* (which enables the Fed to lend directly to "primary dealers," such as Goldman Sachs and Citigroup, which was previously against Fed policy), and
- *Commercial Paper Funding Facility* (which makes the Fed a crucial source of credit for non-financial businesses in addition to commercial banks and investment firms).

Finally, while remaining the supervisor of 5,000 US bank holding companies and 830 state banks, the Fed has taken on substantial new regulatory powers. Under the Wall Street Reform and Consumer Protection Act, known as the Dodd-Frank law (signed July 21, 2010), the central bank gains the authority to control the lending and risk taking of the largest, most "systemically important" banks, including investment banks Goldman Sachs Group and Morgan Stanley, which became bank holding companies in September 2008. The Fed also gains authority over about

440 thrift holding companies and will regulate "systemically important" nonbank financial firms, including the biggest insurance companies, Warren Buffett's Berkshire Hathaway Inc., and General Electric Capital Corp. It is also now required to administer "stress tests" at the biggest banks every year to determine whether they need to set aside more capital. The law prescribes that the largest banks write "living wills," approved by the Fed, that will make it easier for the government to break them up and sell the pieces if they suffer a Lehman Brothers-style meltdown. The Fed also houses and funds a new federal consumer protection agency (headed on an interim basis, as of September 2010, by Elizabeth Warren), which operates independently.

All of this makes the Federal Reserve a far more powerful actor within the US economy. The justification put forward is that without the Fed's bold actions the result would have been utter financial catastrophe, and that with its new powers and functions the institution will be better able to prevent future economic crises. Critics say that catastrophe has merely been delayed.[28]

Actions by Other Nations and Their Central Banks

In November 2008, China announced a stimulus package totaling 4 trillion yuan ($586 billion) as an attempt to minimize the impact of the global financial crisis on its domestic economy. In proportion to the size of China's economy, this was a much larger stimulus package than that of the US. Public infrastructure development made up the largest portion, nearly 38 percent, followed by earthquake reconstruction, funding for social welfare plans, rural development, and technology advancement programs. The stimulus program was judged a success, as China's economy (according to official estimates) continued to expand, though at first at a slower pace, even as many other nations saw their economies contract.

In December 2009, Japan's government approved a stimulus package amounting to 7.2 trillion yen ($82 billion), intended to stimulate employment, incentivize energy-efficient products, and support business owners.

Europe also instituted stimulus packages: in November 2008, the European Commission proposed a plan for member nations amounting to 200 billion euros including incentives to investment, lower interest rates, tax rebates (notably on green technology and eco-friendly cars), and social measures such as increased unemployment benefits. In addition, individual European nations implemented plans ranging in size from 0.6 percent of GDP (Italy) to 3.7 percent (Spain).

The European Central Bank's response to sovereign debt crises, primarily affecting Greece and Ireland but likely to spread to Spain and Portugal, has included a comprehensive rescue package (approved in May 2010) worth almost a trillion dollars. This was accompanied by requirements to cut deficits in the most heavily indebted countries; the resulting austerity programs led, as already noted, to widespread domestic discontent. Greece received a $100 billion bailout, along with a punishing austerity package, in the spring of 2010, while Ireland got the same treatment in November.

A meeting of central bankers in Basel, Switzerland, in September 2010 resulted in an agreement to require banks in the OECD nations to progressively increase their capital reserves starting Jan. 1, 2013. In addition, banks will be required to keep an emergency reserve known as a "conservation buffer" of 2.5 percent. By the end of the decade each bank is expected to have rock-solid reserves amounting to 8.5 percent of its balance sheet. The new rules will strengthen banks against future financial crises, but in the process they will curb lending, making economic recovery more difficult.

After All the Arrows Have Flown

What's the bottom line on all these stimulus and bailout efforts? In the US, $12 trillion of total household net worth disappeared in 2008, and there will likely be more losses ahead, largely as a result of the continued fall in real estate values, though increasingly as a result of job losses as well. The government's stimulus efforts, totaling less than $1 trillion, cannot hope to make up for this historic evaporation of wealth. While indirect subsidies may temporarily keep home prices from falling dramatically, that just keeps houses less affordable to workers making less income. Meanwhile, the bailouts of banks and shadow banks have been characterized as government throwing money at financial problems it cannot solve, rewarding the very people who created them. Rather than being motivated by the suffering of American homeowners or governments in over their heads, the bailouts of Fannie Mae and Freddie Mac in the US, and Greece and Ireland in the EU, were (according to critics) essentially geared toward securing the investments of the banks and the wealthy bond holders.

These are perhaps facile criticisms. It is no doubt true that, without the extraordinary measures undertaken by governments and central banks, the crisis that gripped US financial institutions in the fall of 2008 would have deepened and spread, hurling the entire global economy into a depression surpassing that of the 1930s.

Facile or not, however, the critiques nevertheless contain more than a mote of truth.

The stimulus-bailout efforts of 2008–2009 — which in the US cut interest rates from five percent to zero, spent up the budget deficit to ten percent of GDP, and guaranteed trillions to shore up the financial system — arguably cannot be repeated. In principle, there are ways of conjuring more trillions into existence for such a purpose, as we will see in Chapter 6. However, in Washington the political headwinds against further government borrowing are now gale-force. Thus the realistic likelihood of another huge Congressionally allocated stimulus package is vanishingly small; if more trillions materialize, they are likely to appear in the form of Fed-funded bailouts or quantitative easings. The stimulus-bailout programs constituted quite simply the largest commitments of funds in world history, dwarfing the total amounts spent in all the wars of the 20th century in inflation-adjusted terms (for the US, the cost of World War II amounted to $3.2 trillion). Not only the US, but Japan and the European nations as well may have exhausted their arsenals.

But more will be needed as countries, states, counties, and cities near bankruptcy due to declining tax revenues. Meanwhile, the US has lost 8.4 million jobs — and if loss of hours worked is considered that adds the equivalent of another 3 million; the nation will need to generate an extra 450,000 jobs each month for three years to get back to pre-crisis levels of employment. The only way these problems can be allayed (not fixed) is through more central bank money creation and government spending.

Austrian-School and post-Keynesian economists have contributed a basic insight to the discussion: Once a credit bubble has inflated, the eventual correction (which entails destruction of credit and assets) is of greater magnitude than government's ability to spend. The cycle must sooner or later play itself out.

There may be a few more arrows in the quiver of economic policy makers: central bankers could try to drive down the value of domestic currencies to stimulate exports; the Fed could also engage in more quantitative easing. But these measures will sooner or later merely undermine currencies (we will return to this point in Chapter 6).

Further, the way the Fed at first employed quantitative easing in 2009 was minimally productive. In effect, QE1 (as it has been called) amounted to adding about a trillion dollars to banks' balance sheets, with the assumption that banks would then use this money as a basis for making loans.[29] The "multiplier effect" (in which banks make loans in amounts many times the size of deposits) should theoretically have resulted in the creation of roughly $9 trillion within the economy. However, this did not happen: because there was reduced demand for loans (companies didn't want to expand in a recession and families didn't want to take on more debt), the banks just sat on this extra capital. Perhaps a better result could have been obtained if the Fed were somehow to have distributed the same amount of money directly to debtors, rather than to banks, because then at least the money would either have circulated to pay for necessities, or helped to reduce the general debt overhang. But this would require actions far removed from the Fed's mandate.

In November 2010, the Fed again resorted to quantitative easing ("QE2"). This time, instead of purchasing mortgage securities, thus inflating banks' balance sheets, the Fed set out to purchase Treasuries — $600 billion worth, in monthly installments lasting through June 2011. While QE1 was essentially about saving the banks, QE2 was about funding Federal government debt interest-free. Because the Federal Reserve rebates its profits (after deducting expenses) to the Treasury, creating money to buy government debt obligations is an effective way of increasing that debt without increasing interest payments. Critics describe this as the government "printing money" and assert that it is highly inflationary; however, given the extremely deflationary context (trillions of dollars' worth of write-downs in collateral and credit), the Fed would have to "print" far more than it is doing to result in serious inflation. Nevertheless, as we will see in Chapter 5 in a discussion of "currency wars," other nations view this strategy as a way to drive down the value of the dollar so as to decrease the value of foreign-held dollar-denominated debt — in effect forcing other nations to pay for America's financial folly.

In any case, the Federal Reserve has effectively become a different institution since the crisis began. It and certain other central banks have taken on most of the financial bailout burden (dealing in trillions rather than mere hundreds of billions of dollars) simply because they have the power to create money with which to guarantee banks against losses and buy government debt. Together, central banks and governments are barely keeping the wheels on the economy, but their actions come with severe long-term costs and risks. And what they can actually accomplish is most likely limited anyway. Perhaps the situation is best summed up in a comment from a participant at the central bankers' annual gathering in Jackson Hole, Wyoming in August 2010: "We can't create growth ourselves, all we can do is create the conditions that make growth possible."[30]

BOX 2.3 Just a Little Sideshow

The big banks that were involved in securitizing mortgages and trading them in bundles during the past 15 years purposefully evaded local legal requirements for registering mortgages with a county recorder of deeds as they changed hands. Nor did the banks bother to transfer to the buyer a proper document of assignment evidencing the sale. Mortgages were bundled up into trusts for the purpose of securitizing them to investors, but the trusts were also never given proper legal evidence of the assignment of the mortgages.

Then, when the housing market crashed and banks began millions of foreclosure proceedings, they created the assignments after the fact, using "robo-signers" to submit legal documents to the courts (in one such case the signer had been dead for over five years) and falsified notarizations. In thousands of documented cases foreclosures were conducted even though the borrower was not notified in advance, or the borrower was told by the bank to withhold payments in order to qualify for a mortgage modification but then declared in default by the bank, or the bank added thousands of dollars of "late fees" to the borrower's account, forcing the borrower into default.

In a landmark ruling in January 2011, the Massachusetts Supreme Court held that two banks foreclosed wrongly on two homeowners using suspect paperwork. Attorneys General in 50 states are investigating banks' foreclosure processes. Many observers are questioning whether the banks actually technically own hundreds of billions of dollars' worth of securitized mortgage assets on their balance sheets. If further court rulings go against the banks, the result could be fatal for several "too-big-to-fail" institutions.

Investors who bought MBSs are filing fraud claims against the banks, arguing that these securities were never properly collateralized. Their claims against the banks could amount to trillions of dollars.

The Federal government is implicated as well. Fannie Mae and Freddie Mac now face much higher losses on their portfolios of trillions of dollars' worth of home mortgages, and will therefore likely have to turn to the government for further capital infusions.

L. Randall Wray, a Professor of Economics at the University of Missouri, Kansas City, claims that most mortgage-backed securities are in reality not backed by anything, since the electronic securitization process that most banks used operated illegally. According to Wray, lenders may have the right to foreclose in some instances, but only if they have a clear record of each sale of the mortgage—but electronic securitization in most instances destroyed those records.[31]

The new Congress is likely to try to find a way for the banks to escape this mess, perhaps by simply writing a law declaring the mortgages in question to be valid even without proper documentation. But it is doubtful whether such a law would hold up to scrutiny by the courts. In the end, it may be up to the Supreme Court to decide on the validity of mortgage claims worth trillions.

Deflation or Inflation?

If the bailouts and stimulus packages are effectively just a way of buying time, then there is further trouble ahead—but trouble of what sort? Typically, financial crises play out as inflation or deflation. There is considerable controversy among forecasters as to which will ensue. Let's examine the arguments.

The Inflation Argument

Many economic observers (especially the hard money advocates) point out that the amount of debt that many governments have taken on cannot realistically be repaid, and that the US government in particular will have great difficulty fulfilling its obligations to an aging citizenry via programs like Social Security, Medicare, and Medicaid. The only way out of the dilemma — and it is a time-tested if dangerous strategy — is to inflate the currency. The risk is that inflation undermines the value of the currency and wipes out savings.[32]

There are many fairly recent historic examples, as well as ancient ones going back to the very earliest days of money. The Romans generated inflation by debasing their coinage — gradually reducing the precious-metal content until coins were almost entirely made of base metals. With the advent of paper money, currency inflation became much easier and more tempting: Germany famously inflated away its onerous World War I reparations burdens during the early 1920s. Between June and December 1922, Germans' cost of living increased approximately 1,600 percent, and citizens resorted to carrying bundles of banknotes in wheelbarrows merely to purchase daily necessities; some even used currency as wallpaper. In the United States, hyperinflation occurred during the Revolutionary War and the Civil War. Hungary inflated its currency at the end of World War II, as did Yugoslavia in the late 1980s just before breakup of the country. During the 2000s, Zimbabwe inflated its currency so dramatically that eventually banknotes were being circulated with a face value of 100 trillion Zimbabwe dollars. In each case the result has been the same: a complete gutting of savings and an eventual re-valuation of the currency — in effect, re-setting the value of money from scratch.

How does a nation inflate its currency? There are two primary routes: maintaining very low interest rates encourages borrowing (which, with fractional reserve banking, results in the creation of more money); or direct injection by government or central banks of new money into the economy. This in turn can happen via the central bank creating money with which to buy government debt, or by government creating money and distributing it either to financial institutions (so they can make more loans) or directly to businesses and citizens.

Those who say we are heading toward hyperinflation argue either that existing bailouts and stimulus actions by governments and central banks are inherently inflationary; or that, if the economy relapses, the Federal Reserve will create fresh money not only to buy government debt, but to bail out financial institutions once again. The addition of all this new money, chasing after a limited pool of goods and services, will inevitably cause the currency to lose value.[33]

The Deflation Argument

Others say the most likely course for the world economy is toward continued deleveraging by businesses and households, and this ongoing shedding of debt (mostly through defaults and bankruptcies) will exceed either the ability or willingness of governments and central banks to inflate the currency, at least over the near-term (the next few years). In this view, those who see government actions so far as inflationary fail to see that all that the expansion of public debt has accomplished is to replace a portion of the amount of private debt that has vanished through deleveraging; total debt has actually declined, even in the face of massive government borrowing.[34]

If a bubble consists of lots of people simultaneously taking advantage of what looks like a once-in-a-lifetime opportunity to get rich quick, deflation is lots of people simultaneously doing what appears to be perfectly sensible (under a different set of circumstances) — saving, paying off debts, walking away from underwater homes, and pulling back on borrowing and spending. The net effect of deflation is the destruction of businesses, the layoff of millions of workers, a drop in consumption levels, and consequently further bankruptcies of businesses due to insufficient purchases of overabundant goods and services.

Deflation represents a disappearance of credit and money, so that whatever money remains has increased purchasing power. Once the bubble began to burst back in 2007–2008, say the deflationists, a process of contraction began that inevitably must continue to the point where debt service is manageable and prices for assets such as homes and stocks are compelling based on long-term historical trends.[35]

Many deflationists tend to agree that the inflationists are probably right in the long run: At some point, perhaps several years from now, some future US administration will resort to truly extraordinary means to avoid defaulting on interest payments on its ballooning debt, as well as to avert social disintegration and restart economic activity.[36]

The Bridge to Nowhere

In general, what we are actually seeing so far is neither dramatic deflation nor hyperinflation. Despite the evaporation of trillions of dollars in wealth during the past four years, and despite government and central bank interventions with a potential nameplate value also running in the trillions of dollars, prices (which most economists regard as the signal of inflation or deflation) have remained fairly stable. (While at the time of this writing food and oil prices are soaring, this is due not to monetary policy but to weather events on one hand, and political turmoil in petroleum exporting nations on the other.) That is not to say that the economy is doing well: the ongoing problems of unemployment, declining tax revenues, and business and bank failures are obvious to everyone (see Box I.1 in the Introduction, "But Isn't the US Economy Recovering?"). Rather, what seems to be happening is that the efforts of the US Federal government and the Federal Reserve have temporarily more or less succeeded in balancing out the otherwise massively deflationary impacts of defaults, bankruptcies, and falling property values. With its new functions, the Fed is acting as the commercial bank of last resort, transferring debt (mostly in the form of MBSs and Treasuries) from the private sector to the public sector. The Fed's zero-interest-rate policy has given a huge hidden subsidy to banks by allowing them to borrow Fed money for nothing and then lend it to the government at a 3 percent interest rate. But this is still not inflationary, because the Federal Reserve is merely picking up the slack left by the collapse of credit in the private sector. In effect, the nation's government and its central bank are together becoming the lender of last resort and the borrower of last resort — and (via the military) increasingly also both the consumer of last resort and the employer of last resort.

How can the US continue to run up deficits at a sizeable proportion of GDP? If other nations did the same, the result would be currency devaluation and inflation. America can get away with it for now because the dollar is the reserve currency of the world, and so if the dollar entirely failed most or all of the global economy would go down with it. Other nations are willing to continue holding dollar-denominated debt obligations simply because they see no better alternative. Meanwhile some currency devaluation actually works to America's advantage by making its exports more attractively priced.

Over the short to medium term, then, the US — and, by extension, most of the rest of the world — appears to have achieved a kind of tentative and painful balance. The means used will prove unsustainable, and in any case this period will be characterized by high unemployment, declining wages, and political unease. While leaders will make every effort to portray this as a gradual return to growth, in fact the economy will remain fragile, highly vulnerable to upsetting events that could take any of a hundred forms — including international conflict, popular unrest and dissent, terrorism, the bankruptcy of a large corporation or megabank, a sovereign debt event (such as a default by one of the European countries now lined up for bailouts), a food crisis, an energy shortage, an environmental disaster, a curtailment of government intervention based on the political shift in the makeup of Congress, or a currency war (again, more on that in Chapter 5).

What should be done to avert further deterioration of the global financial system? Once again, the pubic debate (such as it is) is dominated by the opposed viewpoints of the Keynesians and the Chicago Schoolers — which are approximately reflected in the positions of the US Democratic and Republican political parties.

The Keynesians still see the world through the lens of the Great Depression. During the 1930s, industrialized countries were in the early stages of their shift from an agrarian, coal-based, rural economy to an electrified, oil-based, urban economy — a shift that required enormous infrastructure investments (in new highways, airports, dams, and power lines) that would ultimately pay off handsomely for a nation on the verge of realizing a consumer utopia. All that was needed to initiate the building of that infrastructure was credit — grease for the wheels of commerce. Government got those wheels rolling by taking on debt, with private companies increasingly taking the lead after World War II. The expansion that occurred from the 1950s through 2000, as that infrastructure was built and put to use, easily justified the government pump-priming that initiated the process. Future payments of interest on the government debt could be ensured through growth of the tax base.

Now is different. As we will see in the next two chapters, both the US and the world as a whole have passed a fundamental crossroads characterized by increasing scarcity of energy and crucial minerals. Because of this, strategies of growth that worked reliably in the mid-to-late 20th century — via various forms of business and technological development — have reached a point of diminishing returns.

Thus the Keynesian spending bridge today leads nowhere.

But stopping its construction now will result in a catastrophic weakening of the entire economy. The backstop provided by government spending and central bank debt acquisition is the only thing keeping the system from hurtling into a deflationary spiral. Fiscal conservatives who rail against bigger government and more government debt need to comprehend the alternative — a gaping, yawning economic void. For a mere glimpse of what major government spending cutbacks might look like in the US, consider the impacts on European nations that are being subjected to fiscal austerity measures as a corrective for too-rosy expectations of future growth. The picture is bleak: rising poverty, disappearing social services, and general strikes and protests.

Extreme social unrest could be an eventual result of the gross injustice of requiring a majority of the population to forego promised entitlements and economic relief following the bailout of a small super-wealthy minority on Wall Street. Political opportunists can be counted on to exacerbate that unrest and channel it in ways utterly at odds with society's long-term best interests. This is a toxic brew with disturbing precedents in recent European history.

If the Keynesian remedy doesn't cure the ailment but merely extends the suffering (while increasing government debt to truly toxic levels), the medicine of austerity may have such severe side effects that it could kill the patient outright. Both sides — left and right, the socialists and free-marketers — assume and hope to the point of desperation that their prescription will result in a rapid return to continuous economic growth and low unemployment. But as we are about to see, that hope is futile.

There is no "silver bullet," no magic solution that will turn back the clock to an era of abundant resources and easy growth. For now, all that governments can do is buy time through further deficit spending — ideally, using that time to build infrastructure that will continue to function in the coming era of reduced flows of energy and resources. Meanwhile, we must all find ways to come out from under a burden of debt that will otherwise crush us. The inherent contradiction within this prescription is obvious and unavoidable.

BOX 2.4   Credit: The Economic Magnifier

Credit has a history that goes back almost to the beginnings of civilization. For example, early banks (like the Bardi and Peruzzi banks of the tenth and thirteenth centuries) extended credit to monarchs so the latter could afford to go to war. But, during the past century, the extension of credit has become an overwhelmingly pervasive practice that reaches not just into every government and business, but nearly every household in the industrialized world.

Why this vast, recent expansion of credit? One word sums it up: Growth.

Credit gives us the ability to consume now and pay later. It is an expression of belief on the part of both borrower and lender that *later* the borrower will have a surplus with which to repay *today's* new debt, with interest, while still covering basic operating expenses. We will be better off in the future than we are today.

Modern economic theory treats debt as a neutral transfer between saver and consumer. In a world at the end of growth, it becomes anything other than neutral — as the 'savers' will never be able to obtain their deferred consumption.

In an economy of fixed size, where some enterprises are expanding while others are contracting, credit can play a useful but limited role. In a growing economy, credit finds and creates fabulous new opportunities. If credit expands to an unrealistic degree, or if a formerly growing economy enters a recession, the result can be a credit bubble or debt overhang, leading to widespread debt defaults and a dramatic contraction of credit.

In a serious recession, the economy can suffer a powerful, overwhelmingly debilitating one-two punch. The first comes from the interruption of growth; this in itself dashes hopes and leads to increased unemployment and declining earnings. The second, which is potentially far more damaging, comes from the contraction of credit. During the economic ascent, credit provided fuel and encouragement; on the way down, it steepens the fall and removes safety nets. The collapse of credit can turn an economic pothole into a pit of quicksand.

The end of growth is the ultimate credit event, as everyone gradually comes to realize there will be no surplus *later* with which to repay interest on debt that is accruing *now*.

3 EARTH'S LIMITS: WHY GROWTH WON'T RETURN

The 2008 crude oil price, $147 per barrel, shattered the global economy. The "invisible hand" of economics became the invisible fist, pounding down world economic growth to match the limitations of crude oil production.

— Kenneth Deffeyes (petroleum geologist)

We have just seen why, since 2007, growth has languished for reasons internal to the world financial system — the system of money and debt.

Problems arising from speculative overreach, real estate bubbles, and the inherent Ponzi dynamics of our global debt-based financial structures are endemic and profound. Still, if these were our only difficulties, we might reasonably expect that eventually, once they are sorted out (however painful the process may be), growth will return.

Indeed, that is what nearly everyone assumes. It's a matter of "when," not "if" growth resumes.

But there are seldom-acknowledged factors *external* to financial and monetary systems that are effectively choking off efforts to restart growth. These factors, whose impacts are worsening over time, were briefly alluded to in the Introduction; here we will unpack them in more detail, discussing limits to oil and other energy sources, as well as to food, water, and minerals. We will also explore the increasing cost of industrial accidents and environmental disasters — and why, in the wide wake of global climate change, those costs are likely to escalate to the point that disaster avoidance and recovery will constitute a major portion of future government and private spending. Along the way, we will examine how markets respond to resource scarcity (it's not a clear-cut matter of incrementally rising prices).

Crucially, in this chapter we will see how and why the most important of these non-financial limits to economic expansion are matters of concern not just for future generations, but for markets and policy makers — indeed, for everyone — *today*.

Oil

In the Introduction we briefly surveyed the Peak Oil scenario and the events surrounding the oil price spike of 2008. It is tempting here to launch into a lengthy discussion of Peak Oil and what it means to industrial society. I've been writing about this subject for over a decade, and it would be easy to fill the space between these covers simply with updates to existing publications. But that's not what is required here; for our immediate purposes, all that is needed is an overview of some main points regarding oil depletion that are relevant to the question of whether and how economies can continue growing. Readers who wish to know more about Peak Oil should refer to sources listed in the end notes.[1]

When discussion turns to the economy, most of the ensuing talk tends to focus on money — prices, wages, and interest rates. Yet as important as money is to economies, energy is even more basic. Without energy, nothing happens — quite literally. Energy is not just a commodity; it is the prerequisite for any and all activity. No energy, no economy. (In the next chapter we will examine the argument that we can produce economic growth while using *less* energy — by using energy more efficiently; our conclusion will be that this is possible only to a limited extent and in situations that differ fundamentally from our current one.)

The massive worldwide economic growth of the past two centuries was enabled by humanity's newfound ability to exploit the cheap, abundant energy of fossil fuels. There were of course other factors at work — including division of labor, technological innovation, and increased trade. But if it weren't for oil, coal, and natural gas, we would today all probably be living an essentially agrarian existence similar to that of our 18th-century ancestors — though perhaps with a few additional though minor wind- and water-powered industrial accouterments.

Growth requires not just energy in a general sense, but forms of energy with specific characteristics. After all, the Earth is constantly bathed in energy — indeed, the amount of solar energy that falls on Earth's surface each *hour* is greater than the amount of fossil-fuel energy the world uses every *year*. But sunlight energy is diffuse and difficult to use directly. Economies need sources of energy that are concentrated and controllable, and that can be made to do useful work. From a short-term point of view, fossil fuels proved to be energy sources with highly desirable characteristics: they could be extracted from Earth's crust quite cheaply (at least in the early days), they were portable, and they delivered a lot of energy per unit of weight and/or volume — in most instances, far more than the firewood that people had been accustomed to using.

Oil has the particular advantage of being a liquid, which means that it (and its refined products like gasoline and jet fuel) can easily be stored in tanks and pumped through pipes and hoses. This effectively maximizes portability. As a result, oil has become the basis of world transport systems, and therefore of world trade. If the oil stops flowing, global trade as we know it grinds to a standstill.

The phrase "Peak Oil" is often misunderstood to refer to the total exhaustion of petroleum resources — *running out*. In fact it just signifies the period when the production of oil achieves its maximum rate before beginning its inevitable decline. This peaking and declining of production has already been observed in thousands of individual oilfields and in the total national oil production of many countries including the US, Indonesia, Norway, Great Britain, Oman, and Mexico. Global Peak Oil will certainly occur, of that there can be no doubt. There is still some controversy about the timing of the event: has it already happened, will it occur soon, or can it be delayed for many years or even decades?

In 2010, the International Energy Agency settled the matter. In its authoritative 2010 *World Energy Outlook*, the IEA announced that total annual global crude oil production will probably never surpass its 2006 level.[2] However, the agency fudged the question a bit by declaring that

---

BOX 3.1   **Oil Shock 2011?**

In the early months of 2011 street demonstrations erupted in Iraq, Iran, Tunisia, Egypt, Bahrain, Yemen, Libya, and Algeria. Libya became mired in civil war, and its rate of oil exports fell from 1.3 million barrels per day to a small fraction of that amount. In Saudi Arabia, banned opposition groups threatened a "day of rage." In response to these events, the world oil price—already in the $90 range—shot up to $120. Comparisons with the economic oil price spike of 2008, and its consequences, were inevitable.

Many in the US cheered as decrepit dictators in Egypt and Tunisia fell, and as Gaddafi's hold on Libya seemed to loosen. But as it became apparent that more democracy for North African and Middle Eastern nations would translate to higher gasoline prices for American motorists, the real motives for, and costs of Western nations' decades-long support for autocratic regimes in oil-rich nations became starkly apparent. This was a strategy to enforce "stability" among exporters of the world's most important energy resource, but it was wrong-headed from the start because it could not be sustained on the backs of millions of people with rising expectations but declining ability to afford food and fuel.

If, somehow, serious political disruptions are confined to Libya, Egypt, Tunisia, and Bahrain, oil-importing nations may be able to weather 2011 with minimal GDP declines resulting from $100 oil prices. But it may be only a matter of time until Saudi Arabia is engulfed in sectarian and political turmoil, and when that happens the world will see the highest oil price spike ever, and central banks will be powerless to stop the ensuing economic carnage.

---

the peak was not due to geological constraints, and that total volumes of liquid fuels (including crude oil, biofuels, synthetic oil from tar sands and coal, and natural gas liquids like butane and propane) will continue to grow—just a bit—until 2035. In discussing the IEA report, a few analysts declared that these latter claims were essentially just efforts to avoid panicking the markets.[3]

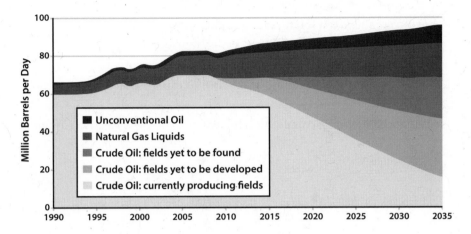

FIGURE 21. Liquid Fuels Forecast, 1990–2035. Source: International Energy Agency, World Energy Outlook 2010.

Scientists who study oil depletion begin with the premise that, for any non-renewable resource such as petroleum, exploration and production proceed on the basis of the best-first or low-hanging fruit principle. Because petroleum geologists began their hunt for oil by searching easily accessible onshore regions of the planet, and because large targets are easier to hit than small ones, the biggest and most conveniently located oilfields tended to be found early in the discovery process.

The largest oilfields — nearly all of which were identified in the decades of the 1930s through the 1960s — were behemoths, each containing billions of barrels of crude and producing oil during their peak years at rates of from hundreds of thousands to several millions of barrels per day. But only a few of these "super-giants" were found. Most of the world's other oilfields, numbering in the thousands, are far smaller, containing a few thousand up to a few millions of barrels of oil and producing it at a rate of anywhere from a few barrels to several thousand barrels per day. As the era of the super-giants passes, it becomes ever more difficult and expensive to make up for their declining production of cheap petroleum with oil from newly discovered oilfields that are smaller and less accessible, and therefore on average more costly to find and develop. As Jeremy Gilbert, former chief petroleum engineer for BP, has put it, "The current fields we are chasing we've known about for a long time in many cases, but they were too complex, too fractured, too difficult to chase. Now our technology and understanding [are] better, which is a good thing, because these difficult fields are all that we have left."[4]

The trends in the oil industry are clear and undisputed: exploration and production are becoming more costly, and are entailing more environmental risks, while competition for access to new prospective regions

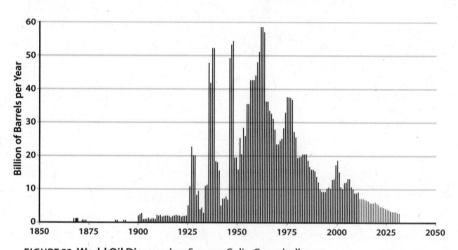

FIGURE 22. **World Oil Discoveries.** Source: Colin Campbell, 2011.

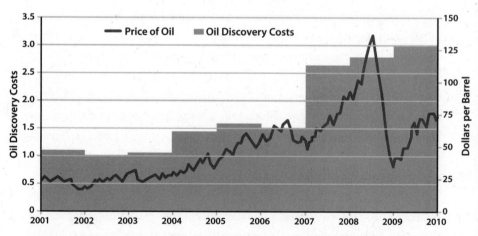

FIGURE 23. **Oil Discovery Costs.** Source: Thomson Reuters, Wood Mackenzie Exploration Service.

is generating increasing geopolitical tension. The rate of oil discoveries on a worldwide basis has been declining since the early 1960s, and most exploration and discovery are now occurring in inhospitable regions such as in ultra-deepwater (at ocean depths of up to three miles) and the Arctic, where operating expenses and environmental risks are extremely high.[5] This is precisely the situation we should expect to see as the low-hanging fruit disappear and global oil production nears its all-time peak in terms of flow rate.

While the US Department of Energy and the IEA continue to produce mildly optimistic forecasts suggesting that global liquid fuels production will continue to grow until at least 2030 or so, these forecasts now come with a semi-hidden caveat: *as long as implausibly immense investments in exploration and production somehow materialize.* This hedged sanguinity is echoed in statements from ExxonMobil and Cambridge Energy Research Associates, as well as a few energy economists. Nevertheless, it is fair to say that most serious analysts now expect a near-term (i.e., within the current decade) commencement of decline in global crude oil *and* liquid fuels production. Prominent oil industry figures such as Charles Maxwell and Boone Pickens say the peak either already has happened or will do so soon.[6] And recent detailed studies by governments and industry groups reached this same conclusion.[7] Toyota, Virgin Airlines, and other major fuel price-sensitive corporations routinely include Peak Oil in their business forecasting models.[8]

Examined closely, the arguments of the Peak Oil naysayers actually boil down to a tortuous effort to say essentially the same things as the Peaksters do, but in less dramatic (some would say *less accurate and useful*) ways. Cornucopian pundits like Daniel Yergin of Cambridge Energy Research Associates speak of a peak not in *supply*, but in *demand* for petroleum (but of course, this reduction in demand is being driven by rising oil prices—so what exactly is the difference?).[9] Or they emphasize that the world is seeing the end of *cheap* oil, not of oil *per se*. They point to enormous and, in some cases, growing petroleum reserves worldwide—yet close examination of these alleged reserves reveals that most consist of "paper reserves" (claimed numbers based on no explicit evidence), or bitumen and other oil-related substances that require special extraction

---

BOX 3.2   **The Mutually Reinforcing Conundrums of Peak Oil and Peak Debt**

The energy returned on the energy invested (EROEI) in producing fossil fuels is declining as we finish picking the low-hanging fruit. According to Charles Hall, who has conducted pioneering studies on "net energy analysis," the EROEI for oil produced in the US was about 100 to one in 1930, declined to 30:1 by 1970, and then to 12:1 by 2005. EROEI figures for coal and gas are also declining, as the easily accessible, high-quality resources are used first.

As EROEI declines over time, an ever-larger proportion of society's energy and resources need to be diverted towards the energy production sector.

Meanwhile, in the social sphere, since money comes into existence via loans but the interest to service those loans isn't created at the same time, the amount of total interest due, society-wide, grows each year.

Thus in the same way that lower EROEI necessitates a shift in investment of energy and other resources from non-energy sectors of the economy towards the energy sector, the interest-servicing requirement of our monetary system diverts more and more resources from the non-finance productive economy towards interest payments.

The result: with time, less of both energy *and* money are available to support basic processes of production and consumption that drove economic growth earlier in the cycle and that support the needs of the population.[13]

---

and processing methods that are slow, expensive, and energy-intensive. Read carefully, the statements of even the most ebullient oil boosters confirm that the world has entered a new era in which we should expect prices of liquid fuels to remain at several times the inflation-adjusted levels of only a few years ago.

Quibbling over the exact meaning of the word "peak," or the exact timing of the event, or what constitutes "oil" is fairly pointless. The oil world has changed. And this powerful shock to the global energy system has just happened to coincide with a seismic shift in the world's economic and financial systems.

The likely consequences of Peak Oil have been explored in numerous books, studies, and reports, and include severe impacts on transport networks, food systems, global trade, and all industries that depend on liquid fuels, chemicals, plastics, and pharmaceuticals.[10] In sum, most of the basic elements of our current way of life will have to adapt or become unsupportable. There is also a strong likelihood of increasing global conflict over remaining oil resources.[11]

Of course, oil production will not cease instantly at the peak, but will decline slowly over several decades; therefore these impacts will appear incrementally and cumulatively, punctuated by intermittent economic and geopolitical crises driven by oil scarcity and price spikes.

Oil importing nations (including the US and most of Europe) will see by far the worst consequences. That's because oil that is available for the export market will dwindle much more quickly than total world oil production, since oil producers will fill domestic demand before servicing foreign buyers, and many oil exporting nations have high rates of domestic demand growth.[12]

Other Energy Sources

Oil is not our only important energy source, nor will its depletion present the only significant challenge to future energy supplies. Coal and natural gas are also pivotal contributors to global energy; they are also fossil fuels, are also finite, and are therefore also subject to the low-hanging fruit principle of extraction. We use these fuels mostly for making electricity, which is just as essential to modern civilization as globe-spanning transport networks. When the electricity goes out, cities go dark, computers blink off, and cash registers fall idle.

As with oil, we are not about to *run out* of either coal or gas. However, here again costs of production are rising, and limits to supply growth are becoming increasingly apparent.[14]

The peak of world coal production may be only years away, as discussed in my 2009 book *Blackout: Coal, Climate and the Last Energy Crisis*. Indeed, one peer-reviewed study published in 2010 concluded that

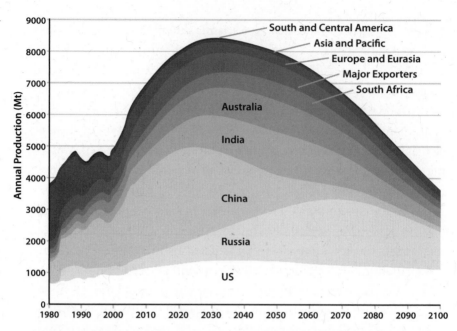

FIGURE 24. World Coal Production Forecast. Source: Energy Watch Group, 2007.

the amount of energy derived from coal globally could peak as early as this year.[15] Some countries that latched onto the coal bandwagon early in the industrial period (such as Britain and Germany) have been watching their production decline for decades. Industrial latecomers are catching up fast by depleting their reserves at phenomenal rates. China, which relies on coal for 70 percent of its energy and has based its feverish economic growth on rapidly growing coal consumption, is now using over 3 billion tons per year—triple the usage rate of the US. Declining domestic Chinese coal production (the national peak will almost certainly occur within the next five to ten years) will lead to more imports, and will therefore put pressure on global supplies.[16] We will explore the implications for China's economy in more detail in Chapter 5.

In the US, most experts still rely on decades-old coal reserves assessments that are commonly (though erroneously) interpreted as indicating that the nation has a 250-year supply. This reliance on outdated and poorly digested data has lulled energy planners, policy makers, and the general public into a dangerous complacency. In terms of the energy it yields, domestic coal production peaked in the late 1990s (more coal is being mined today in raw tonnage, but the coal is of lower and steadily declining energy content). Recent US Geological Survey assessments of some of the most important mining regions show rapid depletion of accessible reserves.[17] No one doubts that there is still an enormous amount of coal in the US, but the idea that the nation can increase total energy production from coal in the years ahead is highly doubtful.

Add to this an exploding Chinese demand for coal imports, and the inevitable result will be steeply rising coal prices globally, even in nations that are currently self-sufficient in the resource. Higher coal prices will in turn torpedo efforts to develop "clean coal" technologies, which on their own are projected to add significantly to the cost of coal-based electricity.[18]

OECD energy demand declined in response to the 2008 financial crisis. If financial turmoil (with resulting reductions in employment and consumption) were to continue in the US and Europe and spread to China, this could help stretch out world coal supplies and keep prices relatively lower. But an economic recovery would quickly lead to much higher energy prices — which in turn would likely force many economies back into recession.

The future of world natural gas supplies is a bit murkier. Conventional natural gas production is declining in many nations, including the US.[19] However, in North America new unconventional production methods based on hydro-fracturing of gas-bearing rocks of low permeability are making significantly larger quantities of gas available, at least over the short term — though at a higher production cost. Due to the temporary supply glut, this higher cost has yet to be reflected in gas prices (currently many companies that specialize in gas "fracking" are subsisting on investment capital rather than profits from production, because natural gas prices are not high enough to make production profitable in most instances).[20] Higher-than-forecast depletion rates add to doubts about whether unconventional gas will be a global game-changer, as it is being called by its boosters, or merely an expensive, short-term, marginal addition to supplies of what will soon be a declining source of energy.[21]

Can other energy sources replace fossil fuels? Some alternatives, such as wind, are seeing rapid growth rates, but still account for only a minuscule share of current global energy supplies. Even if they maintain high

| BOX 3.3 | Applying Time to Energy Analysis |

by Nate Hagens (Excerpted with permission)

Biological organisms, including human societies both with and without market systems, discount distant outputs over those available at the present time based on risks associated with an uncertain future. As the timing of inputs and outputs varies greatly depending on the type of energy, there is a strong case to incorporate time when assessing energy alternatives. For example, the energy output from solar panels or wind power engines, where most investment happens before they begin producing, may need to be assessed differently when compared to most fossil fuel extraction technologies, where a large portion of the energy output comes much sooner, and a larger (relative) portion of inputs is applied during the extraction process, and not upfront. Thus fossil fuels, particularly oil and natural gas, in addition to having energy quality advantages (cost, storability, transportability, etc.) over many renewable technologies, also have a "temporal advantage" after accounting for human behavioral preference for current consumption/return.

In social circumstances where lower discount rates prevail, such as under government mandates and/or in generally more stable societies, longer term energy output becomes more valuable. Less stable societies with higher discount rates will likely handicap longer energy duration investments, as the cost of time will outweigh the value of delayed energy gains. Also in the context of general limits to growth, it is worth noting the evidence that stressed individuals exhibit higher discount rates.

Taking into account time discounting, the EROEI of oil tends to get higher (that is better), while the EROEIs of wind, solar, and corn ethanol tends to get worse. The future energy gain associated with [wind] turbines has decreasing value to users when either (a) the expected lifetime increases or (b) the effective discount rate increases.[27]

If one of the limits to growth consists of limits to capital, then energy sources that tie up capital for a disproportionate length of time before yielding an adequate energy return could be problematic.

rates of growth, they are unlikely to become primary energy sources in any but a small handful of nations by 2050.

In 2009, Post Carbon Institute and the International Forum on Globalization undertook a joint study to analyze 18 energy sources (from oil to tidal power) using 10 criteria (scalability, renewability, energy density, energy returned on energy invested, and so on). While I was the lead author of the ensuing report (*Searching for a Miracle: Net Energy Limits and the Fate of Industrial Societies*), my job was essentially just to synthesize original research and analysis from many energy experts.[22] It was, to my knowledge, the first time so many energy sources had been examined using so many essential criteria. Our conclusion was that there is no credible scenario in which alternative energy sources can entirely make up for fossil fuels as the latter deplete. The overwhelming likelihood is that, by 2100, global society will have *less* energy available for economic purposes, not more.[23]

Here are some relevant passages from that report:

A full replacement of energy currently derived from fossil fuels with energy from alternative sources is probably impossible over the short term; it may be unrealistic to expect it even over longer time frames.... [U]nless energy prices drop in an unprecedented and unforeseeable manner, the world's economy is likely to become increasingly energy-constrained as fossil fuels deplete and are phased out for environmental reasons. It is highly unlikely that the entire world will ever reach an American or even a European level of energy consumption, and even the maintenance of current energy consumption levels will require massive investment.... Fossil fuel supplies will almost surely decline faster than alternatives can be developed to replace them. New sources of energy will in many cases have lower net energy profiles than conventional fossil fuels have historically had, and they will require expensive new infrastructure to overcome problems of intermittency.[24]

Some other studies have reached different, more sanguine conclusions. We believe that this is because they failed to take into account some of the key criteria on which we focused, including the amount of energy returned on the energy that's invested in producing energy (EROEI). Energy sources with a low EROEI cannot be counted as potential primary sources for industrial societies.[25]

As a result of this analysis, we believe that the world has reached immediate, non-negotiable energy limits to growth.[26]

BOX 3.4   **Net Energy**

This "balloon graph" of US energy supplies (Figure 25), developed by Charles Hall of the State University of New York at Syracuse, represents the net energy (vertical axis) and quantity used (horizontal axis) of various energy sources at various times. Arrows show the evolution of domestic oil in terms of EROEI and quantity produced (in 1930, 1970, and 2005), illustrating the historic decline of EROEI for US domestic oil. A similar track for imported oil is also shown. The size of each "balloon" represents the uncertainty associated with EROEI estimates. For example, natural gas has an EROEI estimated at between 10:1 and 20:1 and yields nearly 20 quadrillion Btus (or 20 exajoules). "Total photosynthesis" refers to the total amount of solar energy captured annually by all the green plants in the US including forests, food crops, lawns, etc. (note that the US consumed significantly more than this amount in 2005). The total amount of energy consumed in the US in 2005 was about 100 quadrillion Btus, or 100 exajoules; the average EROEI for all energy provided was between 25:1 and 45:1 (with allowance for uncertainty). The shaded area at the bottom of the graph represents the estimated minimum EROEI required to sustain modern industrial society: Charles Hall suggests 5:1 as a minimum, though the figure may well be higher.[28]

How Markets May Respond to Resource Scarcity: The Goldilocks Syndrome

Before examining limits to non-energy resources, it might be helpful to consider how markets respond to resource scarcity, with petroleum as a highly relevant case in point.

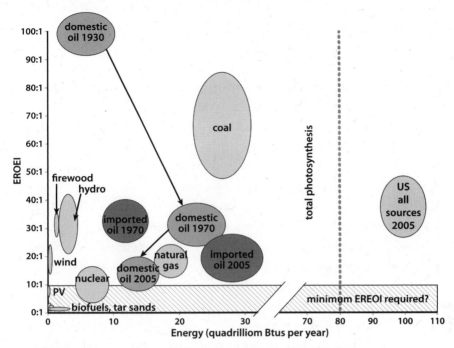

FIGURE 25. Balloon diagram of US energy supplies, including EROEI.

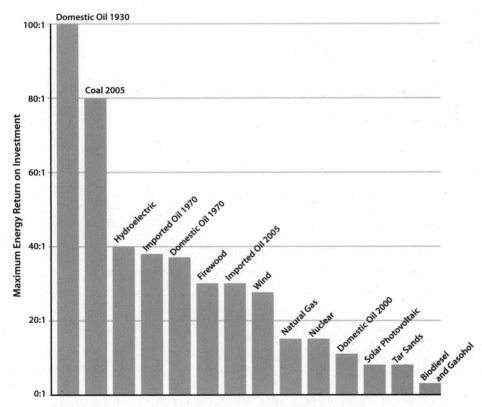

FIGURE 26. Comparison of EROEI for various energy sources. Source: Charles Hall, 2010.

The standard economic assumption is that, as a resource becomes scarce, prices will rise until some other resource that can fill the same need becomes cheaper by comparison. What really happens, when there is no ready substitute, can perhaps best be explained with the help of a little recent history and an old children's story.

Once upon a time (about a dozen years past), oil sold for $20 a barrel in inflation-adjusted figures, and *The Economist* magazine ran a cover story explaining why petroleum prices were set to go much lower.[29] The US Department of Energy and the International Energy Agency were forecasting that, by 2010, oil would probably still be selling for $20 a barrel, but they also considered highly pessimistic scenarios in which the price could rise as high as $30 (those figures are 1996 dollars).[30]

Instead, as the new decade wore on, the price of oil soared relentlessly, reaching levels far higher than the "pessimistic" $30 range. Demand for the resource was growing, especially in China and some oil exporting nations like Saudi Arabia; meanwhile, beginning in 2005, actual world oil production hit a plateau. Seeing a perfect opportunity (a necessary commodity with stagnating supply and growing demand), speculators drove the price up even further.

As prices lofted, oil companies and private investors started funding expensive projects to explore for oil in remote and barely accessible places, or to make synthetic liquid fuels out of lower-grade carbon materials like bitumen, coal, or kerogen.

But then in 2008, just as the price of a barrel of oil reached its all-time high of $147, the economies of the OECD countries crashed. Airlines and trucking companies downsized and motorists stayed home. Demand for oil plummeted. So did oil's price, bottoming out at $32 at the end of 2008.

But with prices this low, investments in hard-to-find oil and hard-to-make substitutes began to look tenuous, so tens of billions of dollars' worth of new energy projects were canceled or delayed. Yet the industry had been counting on those projects to maintain a steady stream of liquid fuels a few years out, so worries about a future supply crunch began to make headlines.[31]

It is the financial returns on their activities that motivate oil companies to make the major investments necessary to find and produce oil. There is a long time lag between investment and return, and so price stability is a necessary condition for further investment.

Here was a conundrum: low prices killed future supply, while high prices killed immediate demand. Only if oil's price stayed reliably within a narrow—and narrowing—"Goldilocks" band could serious problems be avoided. Prices had to stay not too high, not too low—just right—in order to avert economic mayhem.[32]

The gravity of the situation was patently clear. Given oil's pivotal role in the economy, high prices did more than reduce demand, they had helped undermine the economy as a whole in the 1970s and again in 2008. Economist James Hamilton of the University of California, San Diego, has assembled a collection of studies showing a tight correlation between oil price spikes and recessions during the past 50 years. Seeing this correlation, every attentive economist should have forecast a steep recession beginning in 2008, as the oil price soared. "Indeed," writes Hamilton, "the relation could account for the entire downturn of 2007– 08.... If one could have known in advance what happened to oil prices during 2007–08, and if one had used the historically estimated relation [between oil price spikes and economic impacts]...one would have been able to predict the level of real GDP for both of 2008:Q3 and 2008:Q4 quite accurately."[33]

This is not to ignore the roles of too much debt and the exploding real estate bubble in the ongoing global economic meltdown. As we saw in the previous two chapters, the economy was set up to fail regardless of energy prices. But the impact of the collapse of the housing market could only have been amplified by an inability to increase the rate of supply of depleting petroleum. Hamilton again: "At a minimum it is clear that something other than [I would say: "in addition to"] housing deteriorated to turn slow growth into a recession. That something, in my mind, includes the collapse in automobile purchases, slowdown in overall consumption spending, and deteriorating consumer sentiment, in which the oil shock was indisputably a contributing factor."

Moreover, Hamilton notes that there was "an interaction effect be-tween the oil shock and the problems in housing." That is, in many metro-politan areas, house prices in 2007 were still rising in the zip codes closest

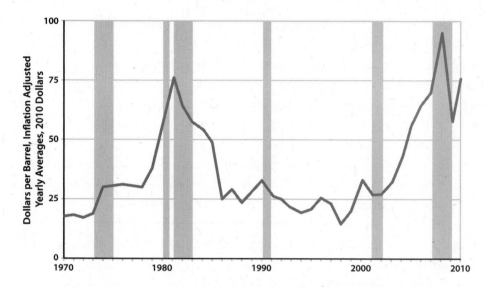

FIGURE 27. Real Oil Prices and Recessions. Rising oil prices bring economic instability. Almost every peak in oil price correlates with an economic downturn. Although the 2000 peak in oil price does not correlate with an official recession, it does correlate with the March 2000 collapse of the dot-com bubble, the unofficial start of the early 2000s Recession. Sources: US Energy Information Administration, US Crude Oil First Purchase Price, The National Bureau of Economic Research.

to urban centers but already falling fast in zip codes where commutes were long.[34]

By mid-2009 the oil price had settled within the "Goldilocks" range — not too high (so as to kill the economy and, with it, fuel demand), and not too low (so as to scare away investment in future energy projects and thus reduce supply). That just-right price band appeared to be between $60 and $80 a barrel.[35]

How long prices can stay in or near the Goldilocks range is anyone's guess (as of this writing, oil is trading in New York for over $100 per barrel), but as declines in production in the world's old super-giant oilfields continue to accelerate and exploration costs continue to mount, the lower boundary of that just-right range will inevitably continue to migrate upward. And while the world economy remains frail, its vulnerability to high energy prices is more pronounced, so that even $80–85 oil could gradually weaken it further, choking off signs of recovery.[36]

> ### BOX 3.5   Declining Energy Intensity
>
> Carey King, a research associate in the University of Texas Center for International Energy and Environmental Policy, in a recent paper in *Environmental Research Letters*, introduced a new measure of energy quality, the Energy Intensity Ratio (EIR).[38] The ratio represents the amount of profit obtained by energy consumers versus energy producers. Higher EIR numbers indicate that more economic value is being derived by households, businesses, and government from each unit of energy consumed.
>
> King plots EIR for various fuels every year since World War II. The resulting graphs show two large declines, one before the recessions of the 1970s and early 1980s, and the other during the 2000s, leading up to the current recession. There have been other recessions in the US since World War II, but the longest and deepest were preceded by sustained declines in EIR for all fossil fuels.
>
> King's analysis suggests that if EIR falls below a certain threshold, the economy ceases growing. For example, in 1972, EIR for gasoline was 5.9 and in 2008 it was 5.5. During times of robust economic growth, such as the 1990s, EIR for gasoline was well over 8.

In other words, oil prices have effectively put a cap on economic recovery.[37] This problem would not exist if the petroleum industry could just get busy and make a lot more oil, so that each unit would be cheaper. But despite its habitual use of the terms "produce" and "production," the industry doesn't *make* oil, it merely *extracts* the stuff from finite stores in the Earth's crust. As we have already seen, the cheap, easy oil is gone. Economic growth is hitting the Peak Oil ceiling.

As we consider other important resources, keep in mind that the same economic phenomenon may play out in these instances as well, though perhaps not as soon or in as dramatic a fashion. Not many resources, when they become scarce, have the capability of choking off economic activity as directly as oil shortages can. But as more and more resources acquire the Goldilocks syndrome, general commodity prices will likely spike and crash repeatedly, making hash of efforts to stabilize the economy.

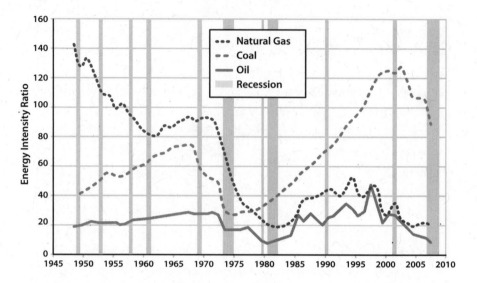

FIGURE 28. EIR declines and recessions. The worst recessions of the last 65 years were preceded by declines in energy quality for oil, natural gas, and coal. Energy quality is plotted using the energy intensity ratio (EIR) for each fuel. Recessions are indicated by gray bars. In layman's terms, EIR measures how much profit is obtained by energy consumers relative to energy producers. The higher the EIR, the more economic value consumers (including businesses, governments, and people) get from their energy. Credit: Carey King.

BOX 3.6	**The Essentials**

Energy, water, and food are all essential and have no substitutes, which means that prices fluctuate wildly in response to small changes in quantity (i.e. demand for them is inelastic). As a side effect of this, their contribution to GNP (price × quantity) increases as their supply declines, which is highly perverse. When financial publications tout "bullish" oil or grain prices, the reader may naturally assume that this constitutes good news. But it's only good for investors in these commodities; for everyone else, higher food and energy prices mean economic pain.

Water

Limits to freshwater could restrict economic growth by impacting society in four primary ways: (1) by increasing mortality and general misery as increasing numbers of people find difficulty filling basic and essential human needs related to drinking, bathing, and cooking; (2) by reducing agricultural output from currently irrigated farmland; (3) by compromising mining and manufacturing processes that require water as an input; and (4) by reducing energy production that requires water. As water becomes scarce, attempts to avert any one of these four impacts will likely make matters worse with regard to at least one of the other three.

There is now widespread concern among experts and responsible agencies that freshwater supplies around the world are being critically overused and degraded, so that water scarcity will increase dramatically as the century wears on. Rivers and streams are being overdrawn, aquifers are being depleted, both surface water and groundwater are being polluted, and sources of flowing surface water — snowpack and glaciers — are receding as a result of climate change.[39]

According to the UN's *Global Environment Outlook 4* (2007), "by 2025, about 1.8 billion people will be living in countries or regions with absolute water scarcity, and two-thirds of the world population could be under conditions of water stress — the threshold for meeting the water requirements for agriculture, industry, domestic purposes, energy and the environment...."[40]

A recent study by a team of researchers at the University of Utrecht and the International Groundwater Resources Assessment Center in Utrecht in the Netherlands estimates that groundwater depletion worldwide went from 99.7 million acre-feet (29.5 cubic miles) in 1960 to 229.4 million acre-feet (55 cubic miles) in 2000.[41] When groundwater is withdrawn and used, it ultimately ends up in the world's oceans, resulting in rising sea levels. However, the contribution of groundwater to sea-level rise will probably diminish in the decades ahead because, in the words of water expert Peter H. Gleick of the Pacific Institute, "as groundwater basins are depleted, there won't be as much water left to send through rain clouds to the oceans."[42]

In the US, the Colorado River — which supplies water to cities such as Phoenix, Tucson, Los Angeles, Las Vegas, and San Diego, as well as providing most of the irrigation water for the Southwest — could be functionally dry within the decade if current trends continue.[43] The snowpack in the headwaters of the Colorado River is decreasing due to climate change and is expected to be at 40 percent below normal in the coming years.

Meanwhile, withdrawals of water continue to increase as population in the region grows. (It is important to distinguish between water withdrawal and water consumption. Water withdrawal represents the total water taken from a source while water consumption represents the amount of that water withdrawal that is not returned to the source, generally lost to evaporation.)[44]

Three billion inhabitants of southern Asia (nearly half the world's population) face a similar crisis: they depend for their water on the great river

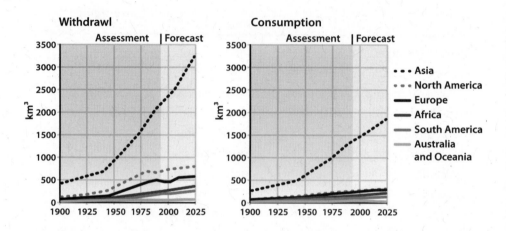

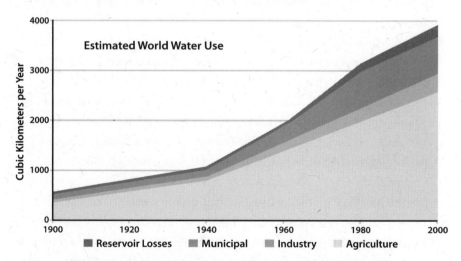

FIGURE 29. World Freshwater Withdrawals and Consumption. Source: United Nations Environment Program (UNEP).

systems that flow from the melting glaciers and snow of the Himalayas — the Ganges, Indus, Brahmaputra, Yangtze, Mekong, Salween, Red River (Asia), Xunjiang, Chao Phraya, Irrawaddy, Amu Darya, Syr Darya, Tarim, and Yellow River. Here again, climate change is reducing the amount of snowpack and shrinking ancient glaciers, while growing populations and expanding economies are making ever-increasing demands on these key waterways.[45]

Life-threatening water shortages have already erupted in parts of Africa. In 2009, Somaliland was gripped by a drought that left thousands of families and their livestock seriously weakened for lack of drinking water. Many water wells dried up altogether, and those that still had water had to serve very large populations, including about 100,000 people displaced by the drought.[46]

Agricultural irrigation accounts for 31 percent of freshwater withdrawals in the US, according to the USGS.[47] The impacts of increasing water shortages on agriculture are illustrated by the dilemma of farmers in California's Central Valley, one of the most productive agricultural areas in America in terms of crop output value per acre. In 2009, in the throes of yet another punishing drought, farmers in Kern County (located in the southern portion of the Central Valley) received less than half their normal water allotment from Federal and state water projects. The local agriculture is highly water-intensive: for Kern County farmers to produce a single orange requires 55 gallons of water, while each peach takes 142 gallons. As a result of the drought, tens of thousands of acres of Kern County farmland were idled.

As snowpack disappears, farmers, ranchers, and cities make up for the loss of running surface water by pumping more from wells. But in many cases this just trades one long-term problem for another: depleting aquifers. The prime example of this trend is the Ogallala aquifer, a vast though shallow underground aquifer located beneath the Great Plains in the United States, which is being drained at an alarming rate. The Ogallala covers an area of approximately 174,000 square miles in portions of eight states (South Dakota, Nebraska, Wyoming, Colorado, Kansas, Oklahoma, New Mexico, and Texas), and supplies water to 27 percent of the irrigated land in the United States.[48] The regions overlying the aquifer are used for ranching and for growing corn, wheat, and soybeans. The Ogallala also provides drinking water to 82 percent of the people who live within the aquifer boundary.[49] Many farmers in the Texas High Plains are already turning away from irrigated agriculture as wells deepen. In most areas covering the aquifer the water table has dropped 10 to 50 feet since groundwater mining began, but drops of over 100 feet have been recorded in several regions.

In the US, only about five percent of freshwater withdrawal is for industrial uses.[50] But these uses support industries that produce, among other things, metals, wood and paper products, chemicals, and gasoline. Industrial water is used for fabricating, processing, washing, diluting, cooling, or transporting products, or for sanitation procedures within manufacturing facilities. Virtually every manufactured product uses water during some part of its production process.

As water becomes scarce, more effort on the part of industry must go toward providing in some other way the same service as cheap water currently provides, almost always at a higher cost. This can mean redesigning industrial processes, or paying more for water brought from further distances.

But moving water takes energy. In California, for example, water pumps use 6.5 percent of the total electricity consumed in the state each year.[51] Desalinating ocean water for industrial, agricultural, and home use also takes energy: the most efficient desalination plants, using reverse osmosis, consume about 2.5 to 3.5 kilowatt hours of energy per cubic meter of fresh water produced.[52]

But if more energy must be used to obtain water as water becomes scarce, more water must be used to obtain energy as energy resources become scarce. Let's return to our earlier example of Kern County, California. In addition to a vital agricultural economy, the county is also host to a $15 billion oil and gas industry—which likewise happens to be very water-intensive. The heavy oil extracted from Kern County oil wells can only flow into and up boreholes when drillers inject enormous amounts of water and steam—320 gallons for every barrel pumped to the surface. Farmers and oil companies must compete for the same dwindling water supplies.[53]

Electricity production requires water, too. About 49 percent of the 410 billion gallons of water the US withdraws daily (if saline water is included) go to cooling thermoelectric power plants, and most of that to cooling coal-burning plants.[54] Nuclear power plants also need substantial amounts of water to cool their reactors. Even the manufacturing of photovoltaic solar panels requires water — in this case, water of exceptionally high purity (though of relatively very small amounts compared to other energy technologies). According to Circle of Blue, a network of journalists and scientists dedicated to water sustainability, "...the competition for water at every stage of the mining, processing, production, shipping and use of energy is growing more fierce, more complex and much more difficult to resolve."[55]

Most nations have been getting steadily more productive with water — that is, water use per unit of GDP has been going up. This is largely due to the shift from agricultural to industrial water use, and also to boosts in efficiency. There is much more that could be done in terms of the latter: water productivity in most sectors could easily double, triple, or more.

Still, across the world conflicts over scarce freshwater resources are multiplying and intensifying. There are many potential flashpoints; for example: a coalition of countries led by Ethiopia is currently challenging old agreements that allow Egypt to use more than half of the Nile's flow. Without the river, all of Egypt would be desert.[56] As users of water, and uses of water, compete for access to dwindling supplies, many nations will find continuing economic growth increasingly put at risk.

By itself, water scarcity is not likely to be an immediate limiting factor for economic growth for the US, at least for the next couple of decades. But it is already a serious problem in many other nations, including much of Africa and most of the Arab world.[57] And water scarcity subtly tightens all the other constraints we are discussing.

Food

In addition to water, people need food for their very existence. Thus food is also essential to economic growth.

Problems with maintenance of far-flung and intensive food production systems played a role in the collapse of previous civilizations, including the Roman Empire.[58] Mesopotamia, the green and lush center of the Sumerian and Babylonian civilizations, was largely turned to desert as a result of soil erosion. The Mayan civilization likewise succumbed to declining food production, according to recent archaeological research.[59]

Industrial societies have skirted what would otherwise have been limiting factors to food production using irrigation, new crop varieties, fertilizers, herbicides and pesticides, and mechanization — as well as expanded transport networks that allow local abundance to be shared globally. In terms of productivity, 20th-century agriculture was an unprecedented success: grain production increased an astounding 500 percent (from under 400 million tons in 1900 to nearly two billion in 2000). This achievement mostly depended on the increasing use of cheap and temporarily abundant fossil fuels.[60]

At the beginning of the 20th century, most people farmed and agriculture was driven by muscle power (animal and human). Today in most countries, farmers make up a much smaller proportion of the population than was formerly the case and agriculture is at least partly mechanized. Fuel-fed machines plow, plant, harvest, sort, process, and deliver foods, and industrial farmers typically work larger parcels of land. They also typically sell their harvest to a distributor or processor, who then sells packaged food products to a wholesaler, who in turn sells these products to chains of supermarkets or restaurants. The ultimate consumer of food is thus several steps removed from the producer, and food systems in most nations or regions have become dominated by a few giant multinational seed companies, agricultural chemicals corporations, and farm machinery manufacturers, as well as food wholesalers, distributors, and supermarket and fast-food chains.

Farm inputs have also changed. A century ago, farmers saved seeds from year to year, while soil amendments were likely to come from the farm itself in the form of animal manures. Farmers only bought basic implements, plus some useful materials such as lubricants. Today's industrial farmer relies on an array of packaged products (seeds, fertilizers, pesticides, herbicides, feed, antibiotics), as well as fuels, powered machines, and spare parts. The annual cash outlays for these can be daunting, requiring farmers to take out substantial loans.

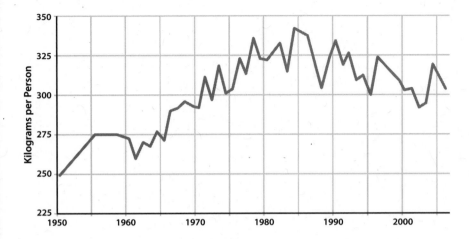

FIGURE 30. World Grain Production Per Person, 1950–2006. Since peaking in 1984, world grain production per person has been falling. Source: Earth Policy Institute.

FIGURE 31. World Grain Stocks, 1960–2010. After peaking in 1986, world stocks have also been declining. Source: Earth Policy Institute.

The path to our current food abundance was littered with incidental costs, most borne by the environment. Agriculture has become the single greatest source of human impact upon the planet as a result of soil salinization, deforestation, loss of habitat and biodiversity, fresh water scarcity, and pesticide pollution of water and soil.[61] Fertilizer use worldwide increased 500 percent from 1960 to 2000, and this contributed to

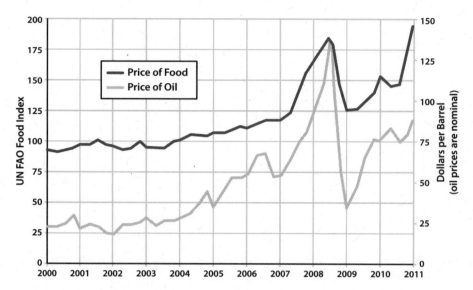

FIGURE 32. World Food and Oil Prices, 2000–2010. There is a strong correlation between food and oil prices. When oil prices spiked in the summer of 2008, food prices reached their highest levels since the UN began collecting food price data. In January 2011, food prices jumped again. This time they reached almost 200 on the FAO's food price index, the highest level ever recorded. Source: UN Food and Agriculture Organization (FAO), US Energy Information Administration.

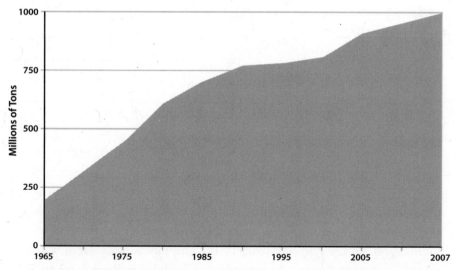

FIGURE 33. World Nitrogen Fertilizer Consumption. Source: UN Food and Agriculture Organization (FAO).

an explosion of "dead zones" in seas and oceans, upsetting a process of nutrient cycling that has existed for billions of years.[62]

There have been some environmental improvements to agriculture in recent years: US farming is more energy efficient than it was a couple of decades ago, fertilizer use has declined somewhat, and more effort goes toward soil conservation. But in general, and especially on the global scene, as food production has grown, so have environmental impacts.[63]

Now, further expansion of the food supply appears problematic. World grain production per capita peaked in 1984 at 342 kg annually. For many years production has not met demand, so the gap has been filled by dipping into carryover stocks; currently, less than two months' supply remains as a buffer.[64]

The challenges to increasing production come from several directions simultaneously: water scarcity (see above), topsoil erosion (we are "mining" topsoil with industrial agriculture at almost four times the rate we are mining coal—over 25 billion tons per year versus 7 billion tons), declining soil fertility, limits to arable land, declining seed diversity, increasing requirements for inputs (pests are developing resistance to common pesticides and herbicides, requiring larger doses), and, not least, increasing costs of fossil fuel inputs.[65]

But as the energy required to run the food system becomes more costly, food is increasingly being used to make energy. Many governments now offer subsidies and other incentives for turning biomass—including food crops—into fuel. This inevitably drives up food prices. Even non-fuel crops such as wheat are affected, as farmers replace wheat fields with more profitable biofuel crops like maize, rapeseed, or soy.

Mineral depletion is also posing a limit to the human food supply. Phosphorus is often a limiting factor in natural ecosystems; that is, the supply of available phosphorus limits the possible size of populations in those environments. That's because phosphorus is one of the three major nutrients required for plant growth (nitrogen and potassium are the other two). Most agricultural phosphorus is obtained from mining phosphate rock: organic farmers use crude phosphate, while conventional industrial farms use chemically treated forms such as superphosphate, triple superphosphate, or ammonium phosphates. Fortunately, phosphorus can be

BOX 3.7   Food Crisis in 2011?

Floods in Australia and Brazil, drought in northern China, and high oil prices are conspiring to drive up food prices in 2011 to record levels. Stockpiling of grain in China and other industrializing countries on expectation of shortages is putting even more upward pressure on prices. As of February, wheat prices were already up nearly to the record levels seen in 2008, and rice orders in China were running at 2 to 4 times usual quantities.

University of Illinois agricultural economist Darrel Good estimated that US corn stocks at the end of the 2010–11 marketing year would total only 745 million bushels. "That projection represents 5.5 percent of projected marketing year consumption. Stocks as a percent of consumption would be the smallest since the record low 5 percent of 1995–96. And 5 percent is considered to be a minimal pipeline supply." Good also noted that combined corn and soybean acreage would have to increase by 6.5 million acres in 2011 to meet anticipated immediate demand while also allowing for a modest rebuilding of inventories. Almost 5 billion bushels of corn will be used in ethanol production this year.[72]

Meanwhile, as Lester Brown, founder of Earth Policy Institute, pointed out in a prominent article ("The Great Food Crisis of 2011"),

"Two huge dust bowls are forming, one across northwest China, western Mongolia, and central Asia; the other in central Africa. Each of these dwarfs the US dust bowl of the 1930s. Satellite images show a steady flow of dust storms leaving these regions, each one typically carrying millions of tons of precious topsoil. In North China, some 24,000 rural villages have been abandoned or partly depopulated as grasslands have been destroyed by overgrazing and as croplands have been inundated by migrating sand dunes."[73]

High food prices were frequently cited as factors contributing to the uprisings in Tunisia and Egypt early in 2011.

recycled, as the Chinese did in their traditional food-agriculture systems, where human and animal wastes were returned to the soil. But today vast amounts of what might otherwise be valuable soil nutrients are flushed down waterways, and wind up being deposited at the mouths of rivers.[66]

In 2007, Canadian physicist and agricultural consultant Patrick Déry studied phosphorus production statistics worldwide using Hubbert linearization analysis (a technique used to forecast oil depletion rates) and concluded that the peak of phosphate production has been passed for both the United States (1988) and for the world as a whole (1989). Déry looked at data not only for phosphate that is currently commercially minable, but for reserves of rock phosphate of lower concentrations; he found — no surprise — that these would be more costly to exploit from economic, energetic, and environmental standpoints.[67] Déry's conclusions are echoed in a recent report by Britain's Soil Association (see Box 3.8 in this chapter, "Does Peak Phosphorus Mean Peak Food?")

There are three main solutions to the problem of Peak Phosphate: composting of human wastes, including urine diversion; more efficient application of fertilizer; and farming in such a way as to make existing soil phosphorus more accessible to plants.

Food supply challenges extend from farms to the world's oceans. Fish like cod, sardines, haddock, and flounder have been favorites for decades in Europe and North America, but many of these species are now endangered. Global marine seafood capture peaked in 1994. An international group of ecologists and economists warned in 2006 that the world will run out of wild seafood by 2048 if steep declines in marine species continue at current rates. They noted that as of 2003, 29 percent of all fished species had collapsed, meaning they were at least 90 percent below their historic maximum catch levels. The rate of population collapses continues to accelerate. The lead author of the group's report, Boris Worm, was quoted as saying, "We really see the end of the line now. It's within our lifetime. Our children will see a world without seafood if we don't change things."[68] According to a more recent study, many types of fish have great difficulty recovering even if over-fishing stops. After 15 years of conservation efforts, many stocks had barely increased in numbers. Cod, for example, failed

---

BOX 3.8   Does Peak Phosphorus Equal Peak Food?

A recent report from the Soil Association of Britain concludes that supplies of phosphate rock are running out faster than previously thought
and that declining supplies and higher prices of phosphate are a new
threat to global food security. *A Rock and a Hard Place: Peak Phosphorus
and the Threat to Our Food Security* highlights the urgent need for farming to become less reliant on phosphate rock-based fertilizer.[74]

Intensive agriculture depends on phosphate to maintain soil fertility.
Globally, 158 million metric ton of phosphate rock is mined each year, but
the mineral is non-renewable and supplies are finite. Recent analysis suggests that the world may hit "peak phosphate" as early as 2033; production in the US is already declining. As with Peak Oil, supplies will become
increasingly scarce and expensive even before production declines, as
mining companies are forced to move to lower-quality deposits.[75]

This critical issue is currently missing from the global food policy
agenda. Without fertilization from phosphorus, wheat yields could
plummet by half in coming decades. Current prices for phosphate rock
are about twice the level in 2006. The Soil Association report notes that,
"When demand for phosphate fertilizer outstripped supply in 2007/08,
the price of rock phosphate rose 800 percent." Phosphate is essential and
non-substitutable; therefore demand is inelastic.

In 2009, 67 percent of rock phosphate was mined in just three countries—China (35 percent), the US (17 percent), and Morocco and Western
Sahara (15 percent). China has now restricted exports, and the US has
stopped exporting the mineral.[76]

---

to recover at all (see Box 3.13 in this chapter, "Atlantic Cod: A Story of
Renewable Resource Depletion").[69]

Here, then is the overall picture: Demand for food is slowly outstrip-
ping supply. Food producers' ability to meet growing needs is increasingly
being strained by rising human populations, falling freshwater supplies,
the rise of biofuels industries, expanding markets within industrializing
nations for more resource-intensive meat and fish-based diets; dwindling

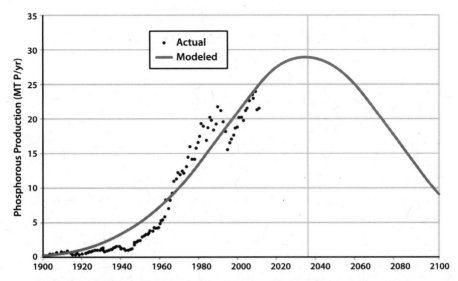

FIGURE 34. World Phosphorus Production, History, and Projection. Source: Cordell et al, 2009, "The Story of Phosphorus: Global Food Security and Food For Thought," *Global Environmental Change* 19:292–305.

wild fisheries; and climate instability. The result will almost inevitably be a worldwide food crisis sometime in the next two or three decades.[70]

The challenges to increasing global food production or even maintaining current rates are linked not only with the other problems discussed in this chapter (changing climate, energy resource depletion, water scarcity, and mineral depletion) but also with the problems discussed in Chapter 2: Modern agriculture requires a system of credit and debt. Unless farmers can obtain credit, they cannot afford increasingly expensive inputs. Food processors and wholesalers likewise require access to credit. Thus a prolonged credit crisis could devastate the world's food supply as dramatically as could any imaginable weather event.

The solution often proposed to these daunting food system challenges is genetic engineering. If we can splice genes to make more productive crop varieties, more nutritious foods, plants that can grow in saltwater, fish that grow faster, or grains that can fix atmospheric nitrogen the way legumes do, then we could reduce the need for freshwater irrigation, nitrogen fertilizers, and overfishing while growing more food and nourishing people better. It sounds too good to be true—and probably is. In reality, most currently patented plant genes merely confer resistance to insect pests or proprietary herbicides; the promise of more nutrient-rich crops and of nitrogen-fixing grains is still years from realization. Meanwhile, the designer-gene seed industry continues to depend on energy-intensive technologies (such as chemical fertilizers and herbicides), as well as centralized production and distribution systems, along with financial systems based on credit and debt. So far, gene splicing in food plants has succeeded mostly in generating enormous profits for an increasingly centralized corporate seed industry, and more debt for farmers. As for gene-altered fish, ecologists warn that while these are meant to be raised in enclosed areas, if even a few accidentally escaped into the wild they could quickly displace remaining related wild populations and upend fragile and already compromised ecosystems.[71]

It's worth noting that for the past few decades a vocal minority of farmers, agricultural scientists, and food system theorists including Wendell Berry, Wes Jackson, Vandana Shiva, Robert Rodale, and Michael Pollan, has argued against centralization, industrialization, and globalization of agriculture, and for an ecological agriculture with minimal fossil fuel inputs. Where their ideas have taken root, the adaptation to Peak Oil and the end of growth will be easier. Unfortunately, their recommendations have not become mainstream, because industrialized, globalized agriculture have proved capable of producing larger short-term profits for banks and agribusiness cartels. Even more unfortunately, the available time for a large-scale, proactive food system transition *before* the impacts of Peak Oil and economic contraction arrive is gone. We've run out the clock.

Metals and Other Minerals

Without metals and a host of other non-renewable minerals, industrial economies could not function. Metals are essential for energy production; for making factory tools, transportation vehicles, and agricultural machinery; and for building the infrastructure of highways, pipes, and power lines that enables modern civilization to function. Hi-tech electronics industries rely on a host of rare metallic and non-metallic minerals ranging from antimony to zinc. All are depleting, and some are already at economically worrisome levels of scarcity.

In principle, there is no sustainable rate of extraction for non-renewable resources: every instance of extraction represents a step toward "running out." During the twentieth century, though, new mining technologies enabled commercially available supplies of most minerals to increase substantially. Ore qualities gradually declined as the low-hanging fruit disappeared, but this trend was countered by the investment of increasing amounts of cheap energy in mining and refining. Globalization also helped, as users of non-renewable resources gained access to virgin deposits in countries where labor costs for mining were minimal. Resource substitution and recycling likewise played their parts in keeping mineral and metal prices low and generally declining.[77]

That price trend seems to have reversed. During the past decade, production rates for many industrially important non-renewable resources have leveled off or, in some cases, begun to decline, while prices have risen.[78] Several recent articles, reports, and studies highlight the predicament of depleting mines, declining ore quality, and rising prices.[79] Data from the US Geological Survey shows that within the US many mineral resources are well past their peak rates of production.[80] These include bauxite (whose production peaked in 1943), copper (1998), iron ore (1951), magnesium (1966), phosphate rock (1980), potash (1967), rare earth metals (1984), tin (1945), titanium (1964), and zinc (1969).[81] As Tom Graedel at Yale University pointed out in a 2006 paper, "Virgin stocks of several metals appear inadequate to sustain the modern 'developed world' quality of life for all of Earth's people under contemporary technology."[82]

The following are just a few examples.

For thousands of years, metal smiths made tools from melted-down "bog iron" (which was mainly composed of an iron-rich ore called goethite), using charcoal as a fuel. In the more recent past iron miners began to extract lower-grade ores such as natural hematite, which were then smelted in coke-fed blast furnaces. Today miners must rely more heavily on taconite, a flint-like ore containing less than 30 percent magnetite and hematite.[83]

According to Julian Phillips, editor of *Gold Forecaster* newsletter, deposits of gold that can be easily mined will probably be exhausted in about 20 years.[84]

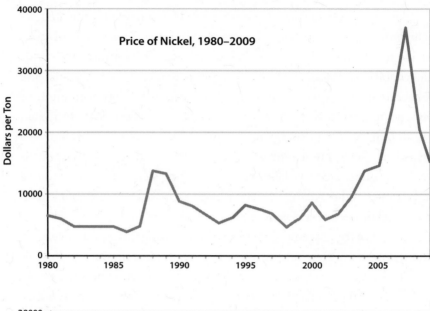

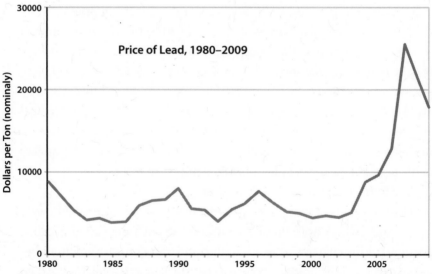

FIGURE 35. Commodity Prices, 1980–2009. Source: UN Conference on Trade and Development (UNCTD).

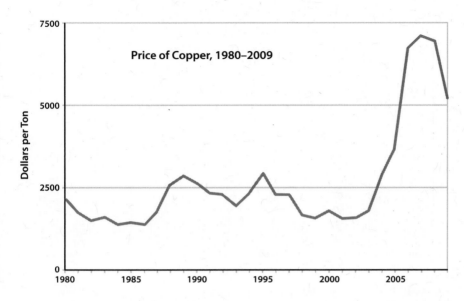

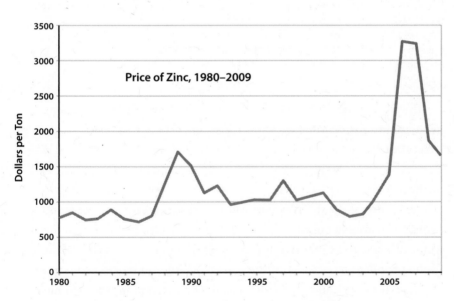

FIGURE 35 (cont'd.)

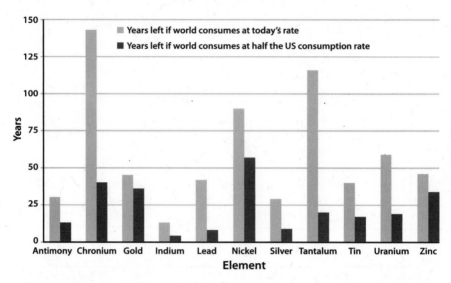

FIGURE 36. Depleting Elements. If nothing changed in our consumption patterns for the above elements, their supply would dwindle relatively slowly (represented in light grey). However, as more countries industrialize, consumption is likely to increase. If the world consumes these resources at only half the current US rate, they will all run out within 60 years (represented in dark grey). Source: Armin Reller of University of Augsburg, Tom Graedel of Yale University, Australian Academy of Science.

There are 17 rare earth elements (REEs) with names like lanthanum, neodymium, europium, and yttrium. They are critical to a variety of high-tech products including catalytic converters, color TV and flat panel displays, permanent magnets, batteries for hybrid and electric vehicles, and medical devices; to manufacturing processes like petroleum refining; and to various defense systems like missiles, jet engines, and satellite components. REEs are even used in making the giant electromagnets in modern wind turbines. But rare earth mines are failing to keep up with demand. China produces 97 percent of the world's REEs, and has issued a series of contradictory public statements about whether, and in what amounts, it intends to continue exporting these elements. The options for other nations, such as the US, are to find substitutes for REEs or to identify new economically viable REE reserves elsewhere in the world.[85]

Indium is used in indium tin oxide, which is a thin-film conductor in flat-panel television screens. Armin Reller, a materials chemist, and his colleagues at the University of Augsburg in Germany, have been investigating the problem of indium depletion. Reller estimates that the world has, at best, 10 years before production begins to decline; known deposits will be exhausted by 2028, so new deposits will have to be found and developed.[86] Some analysts are now suggesting that shortages of energy minerals including indium, REEs, and lithium for electric car batteries could trigger trade wars.[87]

Armin Reller and his colleagues have also looked into gallium supplies. Discovered in 1831, gallium is a blue-white metal with certain unusual properties, including a very low melting point and an unwillingness to oxidize. These make it useful as a coating for optical mirrors, a liquid seal in strongly heated apparatus, and a substitute for mercury in ultraviolet lamps. Gallium is also essential to making liquid-crystal displays in cell phones, flat-screen televisions, and computer monitors. With the explosive profusion of LCD displays in the past decade, supplies of gallium have become critical; Reller projects that by about 2017 existing sources will be exhausted.[88]

Palladium (along with platinum and rhodium) is a primary component in the autocatalysts used in automobiles to reduce exhaust emissions. Palladium is also employed in the production of multi-layer ceramic capacitors in cellular telephones, personal and notebook computers, fax machines, and auto and home electronics. Russian stockpiles have been a key component in world palladium supply for years, but those stockpiles are nearing exhaustion, and prices for the metal have soared as a result.[89]

Uranium is the fuel for nuclear power plants and is also used in nuclear weapons manufacturing; small amounts are employed in the leather and wood industries for stains and dyes, and as mordants of silk or wool. Depleted uranium is used in kinetic energy penetrator weapons and armor plating. In 2006, the Energy Watch Group of Germany studied world uranium supplies and issued a report concluding that, in its most optimistic scenario, the peak of world uranium production will be achieved before 2040. If large numbers of new nuclear power plants are constructed to offset the use of coal as an electricity source, then supplies will peak much sooner.[90]

Tantalum for cell phones. Helium for blimps. The list could go on. Perhaps it is not too much of an exaggeration to say that humanity is in the process of achieving Peak Everything.[91]

> ## BOX 3.9 | Depletion
>
> As a rule of thumb, when the quality of the ore drops, the amount of energy required to extract the resource rises (often the amount of water, too).
>
> Mining companies around the world are reporting declining ore quality, and are using increasing amounts of energy in mining and refining. However, our main sources of energy (fossil fuels) are also depleting non-renewable resources with declining quality. Once energy starts to become scarce and expensive, this sets off a self-reinforcing feedback loop: declining energy supplies make resource extraction more problematic; but since metals and other minerals are essential to energy production, this only makes the energy problem worse, which makes the materials problem worse, which makes the energy problem worse....[92]

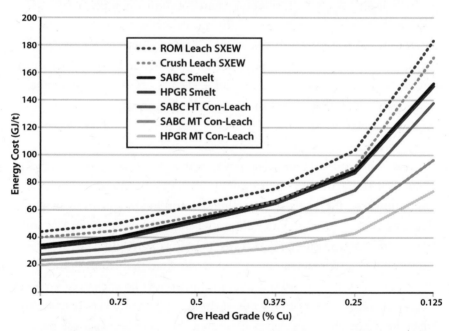

FIGURE 37. Energy Cost of copper production as a function of ore grade. As ore grade declines, energy costs of resource recovery increase. This graph tracks costs of copper production, but the principle could easily be illustrated for other minerals. Source: J. O. Marsden. Hydrometallurgy 2008: Proceedings of the Sixth International Symposium: Energy Efficiency and Copper Hydrometallurgy, Edited by C. A. Young, SME (2008).

Climate Change, Pollution, Accidents, Environmental Decline, and Natural Disasters

Accidents and natural disasters have long histories; therefore it may seem peculiar to think that these could now suddenly become significant factors in choking off economic growth. However, two things have changed.

First, growth in human population and proliferation of urban infrastructure are leading to ever more serious impacts from natural and human-caused disasters. Consider, for example, the magnitude 8.7 to 9.2 earthquake that took place on January 26 of the year 1700 in the Cascadia region of the American northwest. This was one of the most powerful seismic events in recent centuries, but the number of human fatalities, though unrecorded, was probably quite low. If a similar quake were to strike today in the same region — encompassing the cities of Vancouver, Canada; Seattle, Washington; and Portland, Oregon — the cost of damage to homes and commercial buildings, highways, and other infrastructure could reach into the hundreds of billions of dollars, and the human toll might be horrific. Another, less hypothetical, example: the lethality of the 2004 Indian Ocean tsunami, which killed between 200,000 and 300,000 people, was exacerbated by the extreme population density of the low-lying coastal areas of Indonesia, Sri Lanka, and India.

Second, the scale of human influence on the environment today is far beyond anything in the past. In this chapter so far we have considered problems arising from limits to environmental sources of materials useful to society — energy resources, water, and minerals. But there are also limits to the environment's ability to absorb the insults and waste products of civilization, and we are broaching those limits in ways that can produce impacts we cannot contain or mitigate. The billions of tons of carbon dioxide that our species has released into the atmosphere through the combustion of fossil fuels are not only changing the global climate but also causing the oceans to acidify. Indeed, the scale of our collective impact on the planet has grown to such an extent that many scientists contend that Earth has entered a new geologic era — the *Anthropocene*.[93] Humanly generated threats to the environment's ability to support civilization are now capable of overwhelming civilization's ability to adapt and regroup.

Ironically, in many cases natural disasters have actually added to the GDP. This is because of the rebound effect, wherein money is spent on disaster recovery that wouldn't otherwise have been spent. But there is a threshold beyond which recovery becomes problematic: once a disaster is of a certain size or scope, or if conditions for a rebound are not present, then the disaster simply weakens the economy.[94]

Examples of major environmental disasters in 2010 alone include:

- January: a major earthquake in Haiti, with its epicenter 16 miles from the capital Port-au-Prince, left 230,000 people dead, 300,000 injured, and 1,000,000 homeless;
- April–August: the Deepwater Horizon oil rig exploded in the Gulf of Mexico; the subsequent oil spill was the worst environmental disaster in US history;
- May: China's worst floods in over a decade required the evacuation of over 15 million people;
- July–August: Pakistan floods submerged a fifth of the country and killed, injured, or displaced 21 million people, making for the worst natural disaster in southern Asia in decades;
- July–August: Russian wildfires, heat wave, and drought caused hundreds of deaths and the widespread failure of crops, resulting in a curtailing of grain exports; the weather event was the worst in recent Russian history.

But these were only the most spectacular instances. Smaller disasters included:

- February: storms battered Europe; Portuguese floods and mudslides killed 43, while in France at least 51 died;
- April: ash from an Iceland volcano wreaked travel chaos, stranding hundreds of thousands of passengers for days;
- October: a spill of toxic sludge in Hungary destroyed villages and polluted rivers.

This string of calamities continued into early 2011, with deadly, catastrophic floods in Australia, southern Africa, the Philippines, and Brazil.

GDP impacts from the 2010 disasters were substantial. BP's losses from the Deepwater Horizon gusher (which included cleanup costs and

---

BOX 3.10   **The Japan Earthquake**

As this book was in its final stages of preparation for printing, a massive earthquake and tsunami struck northern Japan. Thousands of lives were lost; entire towns were wiped out; nuclear reactors melted down; oil refineries were shuttered; rolling blackouts swept the nation; seaports were seriously damaged; and Toyota, Sony, and other major corporations ceased production.

Only days after these horrific events it was already clear that Japan's economy would be impacted for many months to come. Most oil traders at first assumed that so much destruction in the world's third largest economy would lower world petroleum demand, but others soon argued that Japan would have to import more oil to make up for its lost electrical production capacity, and that refinery outages would put pressure on Asian diesel supplies. Some economists theorized that reconstruction efforts would boost Japan's economy, while others contended that reconstruction could not balance out the enormous GDP hit from lost manufacturing and trading—and that the government, already mired in debt, might not be able to afford to fund complete reconstruction in any case. Energy experts were all generally agreed that the global nuclear power industry had been set back, and that many power plants in the planning stages throughout the world would likely never be built.

The tragic Japanese quake only underscores a general global trend toward ever-higher costs arising from natural disasters and industrial accidents.

---

compensation to commercial fishers) have so far amounted to about $40 billion.[95] The Pakistan floods caused damage estimated at $43 billion, while the financial toll of the Russian wildfires has been pegged at $15 billion.[96] Add in other events listed above, plus more not mentioned, and the total easily tops $150 billion for GDP losses in 2010 resulting from natural disasters and industrial accidents.[97] This does not include costs from ongoing environmental degradation (erosion of topsoil, loss of forests and fish species). How does this figure compare with annual GDP growth? Assuming world annual GDP of $58 trillion and an annual growth rate of three percent, annual GDP growth would amount to $1.74 trillion. There-fore natural disasters and industrial accidents, conservatively estimated, are already costing the equivalent of 8.6 percent of annual GDP growth.

As resource extraction moves from higher-quality to lower-quality ores and deposits, we must expect worse environmental impacts and acci-dents along the way. There are several current or planned extraction proj-ects in remote and/or environmentally sensitive regions that could each result in severe global impacts equaling or even surpassing the Deepwater Horizon blowout. These include oil drilling in the Beaufort and Chukchi Seas; oil drilling in the Arctic National Wildlife Refuge; coal mining in the Utukok River Upland, Arctic Alaska; tar sands production in Alberta; shale oil production in the Rocky Mountains; and mountaintop-removal coal mining in Appalachia.[98]

The future GDP costs of climate change are unknowable, but all in-dications suggest they will be enormous and unprecedented. The most ambitious effort to estimate those costs so far, the *Stern Review on the Economics of Climate Change*, consisted of a 700-page report released for the British government in 2006 by economist Nicholas Stern, chair of the Grantham Research Institute on Climate Change and the Environ-ment at the London School of Economics. The report stated that failure by governments to reduce greenhouse gas emissions would risk caus-ing global GDP growth to lag twenty percent behind what it otherwise might be.[99] The Review also stated that climate change is the greatest and widest-ranging market failure ever seen, presenting a unique challenge for economics.

The Stern Review was almost immediately strongly criticized for un-derestimating the seriousness of climate impacts and the rate at which those impacts will manifest. In April 2008 Stern admitted that, "We un-derestimated the risks...we underestimated the damage associated with temperature increases...and we underestimated the probabilities of tem-perature increases."[100]

The Stern Review is open to criticism not just for its underestimation of climate impacts, but also for its *over*estimation of the ability of alterna-tive energy sources to replace fossil fuels. The report does not take into account EROEI or other aspects of energy quality that are essential to

BOX 3.11   The Environmentalist's Paradox

Environmentalists have long argued that ecological degradation will lead to declines in the general welfare of people who depend on ecosystem services—which in the end includes everyone. Yet the Millennium Ecosystem Assessment found that human well-being has increased despite substantial declines in most global ecosystem services. Were the environmentalists wrong?

A paper by Ciara Raudsepp-Hearne et al., discusses four explanations for these divergent trends: (1) We have measured well-being incorrectly; (2) well-being is dependent on food services, which are increasing, and not on other services, which are declining; (3) technology has decoupled well-being from nature; (4) time lags may lead to future declines in well-being.[109]

Their conclusion: "The environmentalist's paradox is not fully explained by any of the four hypotheses we examined." Despite the assessment that the fourth hypothesis has "weak empirical support," the authors conclude that it may in fact be the best explanation for the divergence in trends. The authors discuss "threshold effects," where ecosystem declines are masked up to a point, but then quickly overwhelm human support systems.

"…[A]nthropogenically driven ecological change has substantial and novel impacts on the biosphere. These changes present new challenges to humanity. The existence of a time lag between the destruction of natural capital and the decline in ecosystem service production provides an explanation of the environmentalist's paradox, but uncertainty about the duration, strength, and generality of this lag prevents us from providing strong support for this hypothesis. However, evidence of past collapses and of declines in natural capital does mean that this hypothesis cannot be rejected."[110]

understanding the economic advantages that fossil fuels have delivered. Since climate is changing mostly because of the burning of fossil fuels, averting climate change is largely a matter of reducing fossil fuel consumption.[101] But as we have seen (and will confirm in more ways in the next chapter), economic growth depends on increasing energy consumption. Due to the inherent characteristics of alternative energy sources, it is extremely unlikely that society can increase its energy production while dramatically curtailing fossil fuel use.[102] Once energy quality factors are taken into account, it is difficult to escape the conclusion that energy substitution will likely be much more expensive than forecast in the Stern Review, and that the price of climate change mitigation — originally estimated at 1 percent of GDP annually in the Review, but later revised to 2 percent — will likely be vastly higher, even ignoring any underestimation of climate change risks and rates.

Another environmental impact that is relatively slow and ongoing and even more difficult to put a price tag on is the decline in the number of non-human species inhabiting our planet. According to one recent study, one in five plant species faces extinction as a result of climate change, deforestation, and urban growth.[103] Many species have existing or potential economically significant uses; the yew tree, for instance, was until recently considered a "trash tree," but is now the source for taxol, relied on by tens of thousands of people as a life-saving treatment for breast, prostate, and ovarian cancers. Sales of the drug have amounted to as much as $1.6 billion in some recent years.[104] As species disappear, potential uses and economic rewards disappear with them.

Another study, this one by the UN, has determined that businesses and insurance companies now see biodiversity loss as presenting a greater risk of financial loss than terrorism — a problem that governments currently spend hundreds of billions of dollars per year to contain or prevent.[105]

Non-human species perform ecosystem services that only indirectly benefit our kind, but in ways that turn out to be crucial. Phytoplankton, for example, are not a direct food source for people, but comprise the base of oceanic food chains — in addition to supplying half of the oxygen produced each year by nature. The abundance of plankton in the world's oceans has declined 40 percent since 1950, according to a recent study,

---

BOX 3.12	Growing Demand for Resources "Threatens EU Economy"

According to a report by the European Environment Agency, released in November 2010, Europe's economy is at risk due to limits of global natural resources. The *Environment State and Outlook Report* said the threat is driven by a need to satisfy rising global consumption.[111]

The report said there were no "quick fixes" and called on businesses, individuals and policymakers to work together to make resource use more efficient.

The report's authors noted that EU environmental policy had "delivered substantial improvements," but added: "major environmental challenges remain, which will have significant consequences for Europe if left unaddressed."

For example, while Europe's network of protected areas and habitats has expanded to cover about 18 percent of the continent's landmass, the EU has failed to meet its target to halt biodiversity loss by 2010.

The report, according to its executive summary, "does not present any warnings of imminent environmental collapse. However, it does note that some local and global thresholds are being crossed, and that negative environmental trends could lead to dramatic and irreversible damage to some of the ecosystems and services that we take for granted."[112]

---

for reasons not entirely clear.[106] This is one of the main explanations for a gradual decline in atmospheric oxygen levels recorded worldwide.[107]

A 2010 study by Pavan Sukhdev, a former banker, to determine a price for the world's environmental assets, concluded that the annual destruction of rainforests entails an ultimate cost to society of $4.5 trillion — $650 for each person on the planet. But that cost is not paid all at once; in fact, over the short term, forest cutting looks like an economic benefit as a result of the freeing up of agricultural land and the production of timber. Like financial debt, deferred environmental costs tend to accumulate until a crisis occurs and systems collapse.[108]

Declining oxygen levels, acidifying oceans, disappearing species, threatened oceanic food chains, changing climate—when considering planetary changes of this magnitude, it may seem that the end of economic growth is hardly the worst of humanity's current problems. However, it is important to remember that we are counting on growth to enable us to solve or respond to environmental crises. With economic growth, we have surplus money with which to protect rainforests, save endangered species, and clean up after industrial accidents. Without economic growth, we are increasingly defenseless against environmental disasters — many of which paradoxically result from growth itself.

Unfortunately, in the case of climate change, there may be a time lag involved (even if we stop carbon emissions today, climate will continue changing for some time due to carbon already in the atmosphere), so that the end of economic growth cannot be counted on to solve the environmental problems that growth has previously generated.

In the Introduction to this book we began with a simple premise: humanity cannot grow consumption and waste streams forever on a finite planet. There are limits. The evidence is clear: We are reaching those limits. It is no longer a matter of saying, "*If* we don't voluntarily bring growth in consumption to an end, *then* we will run into problems." That was a message appropriate to the 1970s or '80s. We didn't change direction then, and now we are nearing or at the point of declining energy, declining freshwater, declining minerals, declining biodiversity…and a declining economy.

Perhaps the meteoric rise of the finance economy in the past couple of decades resulted from a semi-conscious strategy on the part of society's managerial elites to leverage the last possible increments of growth from a physical, resource-based economy that was nearing its capacity. In any case, the implications of the current economic crisis cannot be captured by unemployment statistics and real estate prices. Attempts to restart growth will inevitably collide with natural limits that simply don't respond to stimulus packages or bailouts.[113]

Burgeoning environmental problems require rapidly increasing amounts of effort to fix them. In addition to facing limits on the amount of debt that can be accumulated in order to keep those problems at bay,

---

BOX 3.13 **Atlantic Cod: A Story of Renewable Resource Depletion**

In 1988, Canadian newspaper and magazine articles (exemplified in a prominent *Canadian Geographic* story, "Almighty Cod") celebrated the extraordinary success of the Atlantic Fishery. The fishing industry, including the practice of dragging-trawling, was presented in glowing terms, and it was suggesting that the cod fishery could lift eastern Canada out of its economic doldrums.

Just two years later, scientific reports showed that the fishery was in serious trouble, and fishermen reported their catches were declining rapidly (*Canadian Geographic* ran a new story by the same author titled, "Net Losses: The Sorry State of our Atlantic Fishery"). The turnaround was abrupt and catastrophic.

Canadian media began detailing the horrendous environmental costs of dragging-trawling. And scientists admitted that they really knew very little about cod.

Government quotas (special Company Allotments) succeeded in keeping prices stable, and fishing continued at levels that were still much too high, until it was obvious that cod populations had collapsed.

In 1992, a cod fishing moratorium was declared.

Nearly twenty years have passed since then, and despite the moratorium, fish stocks have not recovered. Indeed there is not much hope that they will at this point. There are still cod out there, but only a few fishermen go after them. The price is still good, but there is simply not enough cod to support the livelihoods of fishermen, let alone an entire industry. So now the industry has shifted its focus to shellfish, shrimp, and lobster.

---

we also face limits to the amounts of energy and materials we can devote to those purposes. Until now the dynamism of growth has enabled us to stay ahead of accumulating environmental costs. As growth ends, the environmental bills for our last two centuries of manic expansion may come due just as our bank account empties.[114]

4 WON'T INNOVATION, SUBSTITUTION, AND EFFICIENCY KEEP US GROWING?

I want to believe in innovation and its possibilities, but I am more thoroughly convinced of entropy. Most of what we do merely creates local upticks in organization in an overall downward sloping curve. In that regard, technology is a bag of tricks that allows us to slow and even reverse the trend, sometimes globally, sometimes only locally, but always only temporarily and at increasing aggregate energy cost.

— Paul Kedrosky (entrepreneur,
editor of the econoblog Infectious Greed)

In the course of researching and writing this book, I discussed its central thesis — that world economic growth has come to an end — with several economists, various businesspeople, a former hedge fund manager, a top-flight business consultant, and the former managing director of one of Wall Street's largest investment banks, as well as several ecologists and environmental activists. The most common reaction (heard as often from the environmentalists as the bankers) was along the lines of: "But capitalism has a few more tricks up its sleeve. It's infinitely creative. Even if we've hit environmental limits to energy or water, the mega-rich will find ways to amass yet more capital on the way down the depletion slope. It'll still look like growth to them."

Most economists would probably agree with the view that environmental constraints and a crisis in the financial world don't add up to the *end* of growth—just a speed bump in the highway of progress. That's because smart people will always be thinking of new technologies and of new ways to do more with less. And these will in turn be the basis of new commercial products and business models.

Talk of limits typically elicits dismissive references to the failed warnings of Thomas Malthus—the 18th-century economist who reasoned that population growth would inevitably (and soon) outpace food production, leading to a general famine. Malthus was obviously wrong, at least in the short run: food production expanded throughout the 19th and 20th centuries to feed a fast-growing population. He failed to foresee the introduction of new hybrid crop varieties, chemical fertilizers, and the development of industrial farm machinery. The implication, whenever Malthus's ghost is summoned, is that *all* claims that environmental limits will overtake growth are likewise wrong, and for similar reasons. New inventions and greater efficiency will always trump looming limit.[1]

In this chapter, we will examine the factors of efficiency, substitution, and innovation critically and see why—while these are key to our efforts to adapt to resource limits—they are incapable of removing those limits, and are themselves subject to the law of diminishing returns. And returns on investments in these strategies are in many instances already quickly diminishing.

Substitutes Forever

It is often said that "the Stone Age didn't end for lack of stones, and the oil age won't end for lack of oil; rather, it will end when we find a cheaper, better source of energy." Variations on that maxim have appeared in ads from ExxonMobil, statements from the Saudi Arabian government, and blogs from pro-growth think tanks—all arguing that the world faces no energy shortages, only energy opportunities.

It's true: the Stone Age ended when our ancient ancestors invented metal tools and found them to be superior to stone tools for certain purposes, not because rocks became scarce. Similarly, in the late-19th century early industrial economies shifted from using whale oil for lubrication and lamp fuel to petroleum, or "rock oil." Whale oil was getting expensive because whales were being hunted to the point that their numbers were dropping precipitously. Petroleum proved not only cheaper and more abundant, it also turned out to have a greater variety of uses. It was a superior substitute for whale oil in almost every respect.

Fast forward to the early 21st century. Now the cheap rock oil is gone. It's time for the next substitute to appear — a magic elixir that will make nasty old petroleum look as obsolete and impractical as whale oil. But what exactly is this "new oil"?

Economic theory is adamant on the point: as a resource becomes scarce, its price will rise until some other resource that can serve the same need becomes cheaper by comparison. That the replacement will prove superior is not required by theory.

Well, there certainly are substitutes for oil, but it's difficult to see any of them as superior — or even equivalent — from a practical, economic point of view.[2]

Just a few years ago, ethanol made from corn was hailed as the answer to our dependence on depleting, climate-changing petroleum. Massive amounts of private and public investment capital were steered toward the ethanol industry. Government mandates to blend ethanol into gasoline further supported the industry's development. But that experiment hasn't turned out well. The corn ethanol industry went through a classic boom-and-bust cycle, and expanding production of the fuel hit barriers that were foreseeable from the very beginning. It takes an enormous land area to produce substantial amounts of ethanol, and this reduces the amount of cropland available for growing food; it increases soil erosion and fertilizer pollution while forcing food prices higher. By 2008, soil scientists and food system analysts were united in opposing further ethanol expansion.[3]

For the market, ethanol proved too expensive to compete with gasoline. But from an energy point of view the biggest problem with corn ethanol was that the amount of energy required to grow the crop, harvest and collect it, and distill it into nearly pure alcohol was perilously close to the amount of energy that the fuel itself would yield when burned in an engine. This meant that ethanol wasn't really much of an energy source at all; making it was just a way of taking existing fuels (petroleum and natural gas) and using them (in the forms of tractor fuel, fertilizer, and fuel for distillation plants) to produce a different fuel that could be used for the same purposes as gasoline. Experts argued back and forth: one critic said the energy balance of corn ethanol was actually negative (less than 1:1) — meaning that ethanol was a losing proposition on a net energy basis.[4] But then a USDA study claimed a positive energy balance of 1.34:1.[5] Other studies yielded slightly varying numbers (the differences had to do with deciding which energy inputs should be included in the analysis).[6] From a broader perspective, this bickering over decimal-place accuracy was pointless: in its heyday, oil had enjoyed an EROEI of 100:1 or more, and it is clear that for an industrial society to function it needs primary energy sources with a minimum EROEI of between 5:1 and 10:1.[7] With an overall societal EROEI of 3:1, for example, roughly a third of all of that society's effort would have to be devoted just to obtaining the energy with which to accomplish all the other things that a society must do (such as manufacture products, carry on trade, transport people and goods, provide education, engage in scientific research, and maintain basic infrastructure). Since even the most optimistic EROEI figure for corn ethanol is significantly below that figure, it is clear that this fuel cannot serve as a primary energy source for an industrial society like the United States.

The problem remains for so-called second- and third-generation biofuels — cellulosic ethanol made from forest and crop wastes and biodiesel squeezed from algae. Extraordinary preliminary claims are being made for the potential scalability and energy balance of these fuels, which so far are still in the experimental stages, but there is a basic reason for skepticism about such claims. With all biofuels we are trying to do something inherently very difficult — replace one fuel, which nature collected and concentrated, with another fuel whose manufacture requires substantial effort on our part to achieve the same result. Oil was produced over the course of tens of millions of years without need for any human work. Ancient sunlight energy was chemically gathered and stored by vast numbers of microscopic aquatic plants, which fell to the bottoms of seas and were buried under sediment and slowly transformed into energy-dense hydrocarbons. All we have had to do was drill down to the oil-bearing rock strata, where the oil itself was often under great pressure so that it flowed easily up to the surface. To make biofuels, we must engage in a variety of activities that require large energy expenditures for growing and fertilizing crops, gathering crops or crop residues, pressing algae to release oils, maintaining and cleaning algae bioreactors, or distilling alcohol to a high level of purity (this is only a partial list). Even with substantial technical advances in each of these areas, it will be impossible to compete with the high level of energy payback that oil enjoyed in its heyday.

This is not to say that biofuels have no future. As petroleum becomes more scarce and expensive we may find it essential to have modest quantities of alternative fuels available for certain purposes even if those alternatives are themselves expensive, in both monetary and energy terms. We will need operational emergency vehicles, agricultural machinery, and some aircraft, even if we have to subsidize them with energy we might ordinarily use for other purposes. In this case, biofuels will not serve as one of our society's primary energy sources — the status that petroleum enjoys today. Indeed, they will not comprise much of an energy *source* at all in the true sense, but will merely serve as a means to transform energy that is already available into fuels that can be used in existing engines in order to accomplish selected essential goals. In other words, biofuels will substitute for oil on an emergency basis, but not in a systemic way.

The view that biofuels are unlikely to fully substitute for oil anytime soon is supported by a recent University of California (Davis) study that concludes, on the basis of market trends only, that "At the current pace of research and development, global oil will run out 90 years before replacement technologies are ready."[8]

It could be objected that we are thinking of substitutes too narrowly. Why insist on maintaining current engine technology and simply switching the fuel? Why not use a different drive train altogether?

Electric cars have been around nearly as long as the automobile itself. Electricity could clearly serve as a substitute for petroleum — at least when it comes to ground transportation (aviation is another story — more on that in a moment). But the fact that electric vehicles have failed for so long to compete with gasoline- and diesel-powered vehicles suggests there may be problems.

In fact, electric cars have advantages as well as disadvantages when compared to fuel-burning cars. The main advantages of electrics are that their energy is used more efficiently (electric motors translate nearly all their energy into motive force, while internal combustion engines are much less efficient), they need less drive-train maintenance, and they are more environmentally benign (even if they're running on coal-derived electricity, they usually entail lower carbon emissions due to their much higher energy efficiency). The drawbacks of electric vehicles are partly to do with the limited ability of batteries to store energy, as compared to conventional liquid fuels. A gallon of gasoline carries 46 megajoules of energy per kilogram, while lithium-ion batteries can store only 0.5 MJ/kg. Improvements are possible, but the ultimate practical limit of chemical energy storage is still only about 6–9 MJ/kg.[9] This is why we'll never see battery-powered airliners: the batteries would be way too heavy to allow planes to get off the ground. This doesn't mean research into electric aircraft should not be pursued: There have been successful experiments with ultra-light solar-powered planes, and electric planes could come in handy in a future where most transport will be by boat, rail, bicycle, or foot. But these will be special-purpose aircraft that can carry only one or two passengers.

The low energy density (by weight) of batteries tends to limit the range of electric cars. This problem can be solved with hybrid power trains—using a gasoline engine to charge the batteries, as in the Chevy Volt, or to push the car directly part of the time, as with the Toyota Prius—but that adds complexity and expense.

So substituting batteries and electricity for petroleum works in some instances, but even in those cases it offers less utility (if it offered *more* utility, we would all already be driving electric cars).[10]

Increasingly, substitution is less economically efficient. But surely, in a pinch, can't we just accept the less-efficient substitute? In emergency or niche applications, yes. But if the less-efficient substitute must replace a resource of profound economic importance (like oil), or if a large number of resources have to be replaced with less-useful substitutes, then the overall result for society is a reduction—perhaps a sharp reduction—in its capacity to achieve economic growth.

As we saw in Chapter 3, in our discussion of the global supply of minerals, when the quality of an ore drops the amount of energy required to extract the resource rises. All over the world mining companies are reporting declining ore quality.[11] So in many if not most cases it is no longer possible to substitute a rare, depleting resource with a more abundant, cheaper resource; instead, the available substitutes are themselves already rare and depleting.

Theoretically, the substitution process can go on forever — as long as we have endless energy with which to obtain the minerals we need from ores of ever-declining quality. But to produce that energy we need more resources. Even if we are using only renewable energy, we need steel for wind turbines and coatings for photovoltaic panels. And to extract *those* resources we need still more energy, which requires more resources, which requires more energy. At every step down the ladder of resource quality, more energy is needed just to keep the resource extraction process going, and less energy is available to serve human needs (which presumably is the point of the exercise).[12]

The issues arising with materials synthesis are similar. In principle it is possible to synthesize oil from almost any organic material. We can make petroleum-like fuels from coal, natural gas, old tires, even garbage. However, doing so can be very costly, and the process can consume more energy than the resulting synthetic oil will deliver as a fuel, unless the material we start with is already similar to oil.

It's not that substitution can never work. Recent years have seen the development of new catalysts in fuel cells to replace depleting, expensive platinum, and new ink-based materials for photovoltaic solar panels that use copper indium gallium diselenide (CIGS) and cadmium telluride to replace single-crystalline silicon. And of course renewable wind, solar, geothermal, and tidal energy sources are being developed and deployed as substitutes for coal.

We will be doing a lot of substituting as the resources we currently rely on deplete. In fact, materials substitution is becoming a primary focus of research and development in many industries. But in the most important cases (including oil), the substitutes will probably be inferior in terms of economic performance, and therefore will not support economic growth.

| BOX 4.1 | Substitution Time Lags and Economic Consequences |

Assume that world oil production peaks this year and begins declining at a rate of two percent per year. We will then need to increase volumes of replacement fuels by this amount plus about 1.5 percent annually in order to fuel a modest rate of economic growth (that's 3.5 percent total). We can theoretically achieve the same amount of growth by increasing transport energy efficiency by 3.5 percent per year, or by pursuing some combination of these two strategies, as long as the total effect is to adjust to declining oil availability while maintaining growth. We will probably not achieve either our substitution or efficiency goals during the first year (it takes time to develop new policies and technologies), so we will face an oil shortfall amounting to 3.5 percent of the total, minus whatever small increment we are able to offset with replacement fuels and efficiency on a short-term basis. The next year will see a similar situation. If, at the moment production started declining, we were smart enough to start investing heavily in substitutes and more efficient transport infrastructure (electric cars and trains), then those investments would start to pay off in three to four years, but it will take even longer—four to five years—for substitution and efficiency to offer significant help.[13]

During these five years, unless we have plans in place to handle fuel shortfalls, adaptation will not be orderly or painless. With a reduction of two percent in oil availability, we may experience a decrease in GDP of three or four percent. Investors will become cautious and job markets will contract. There is no way to know how markets will respond during this period of high insecurity about the future energy supply. The values of currencies, the stock market, bonds, and real estate are all tied to the belief that the economy will grow in the future. With three, four, or five years of recession or depression, belief in future economic growth could wane, causing markets to fall further. The value of many of these assets could fall very significantly. And of course in a recession it may be harder to allocate resources towards innovation.

This is why it is essential to begin investing in efficiency and alternative energy as soon as possible, and choose wisely with regard to those investments.[14]

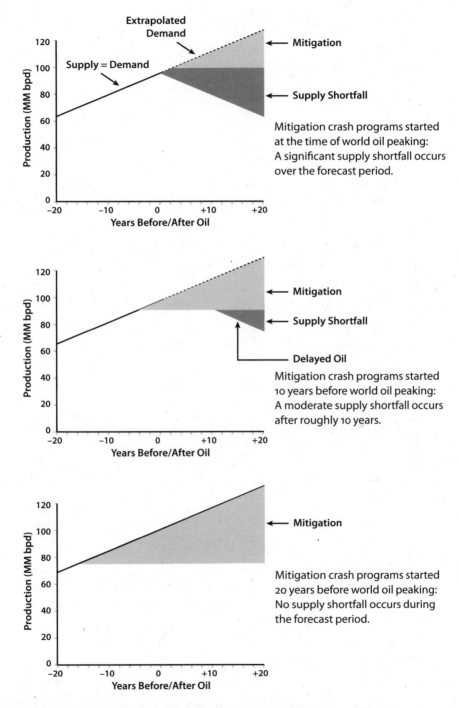

Mitigation crash programs started at the time of world oil peaking: A significant supply shortfall occurs over the forecast period.

Mitigation crash programs started 10 years before world oil peaking: A moderate supply shortfall occurs after roughly 10 years.

Mitigation crash programs started 20 years before world oil peaking: No supply shortfall occurs during the forecast period.

FIGURE 38. The Cost of Delaying Prepatory Response to Peak Oil. Source: Robert L. Hirsch, Roger Bezdek, and Robert Wendling, 2005, "Peaking of World Oil Production: Impacts, Mitigation, & Risk Management."

Energy Efficiency to the Rescue

The historic correlation between economic growth and increased energy consumption is controversial, and I promised in Chapter 3 to return to the question of whether and to what degree it is possible to de-link or decouple the two.

While it is undisputed that, during the past two centuries, both energy use and GDP have grown dramatically, some analysts argue that the causative correlation between energy consumption and growth is not tight, and that energy consumption and economic growth can be decoupled by increasing the efficiency with which energy is used. That is, economic growth can be achieved while using *less* energy.[15]

This has already happened, at least to some degree. According to the US Energy Information Administration,

> From the early 1950s to the early 1970s, US total primary energy consumption and real GDP increased at nearly the same annual rate. During that period, real oil prices remained virtually flat. In contrast, from the mid-1970s to 2008, the relationship between energy consumption and real GDP growth changed, with primary energy consumption growing at less than one-third the previous average rate and real GDP growth continuing to grow at its historical rate. The decoupling of real GDP growth from energy consumption growth led to a decline in energy intensity that averaged 2.8 percent per year from 1973 to 2008.[16]

Translation: We're saved! We just need to double down on whatever we've been doing since 1973 that led to this decline in the amount of energy it took to produce GDP growth.[17]

However, several analysts have pointed out that the decoupling trend of the past 40 years conceals some explanatory factors that undercut any realistic expectation that energy use and economic growth can diverge much further.[18]

One such factor is the efficiency gained through fuel switching. Not all energy is created equal, and it's possible to derive economic benefits from changing energy sources while still using the same amount of total energy. Often energy is measured purely by its heating value, and if one considers only this metric then a British Thermal Unit (Btu) of oil is by definition equivalent to a Btu of coal, electricity, or firewood. But for practical economic purposes, every energy source has a unique profile of advantages and disadvantages based on factors like energy density, portability, and cost of production. The relative prices we pay for natural gas, coal, oil, and electricity reflect the differing economic usefulness of these sources: a Btu of coal usually costs less than a Btu of natural gas, which is cheaper than a Btu of oil, which is cheaper than a Btu of electricity.

Electricity is a relatively expensive form of energy but it is very convenient to use (try running your computer directly on coal!). Electricity can be delivered to wall outlets in billions of rooms throughout the world, enabling consumers easily to operate a fantastic array of gadgets, from toasters and blenders to iPad chargers. With electricity, factory owners can run computerized monitoring devices that maximize the efficiency of automated assembly lines. Further, electric motors can be highly efficient at translating energy into work. Compared to other energy sources, electricity gives us more economic bang for each Btu expended (for stationary as opposed to mobile applications).

As a result of technology developments and changes in energy prices, the US and several other industrial nations have altered the ways they have used primary fuels over the past few decades. And, as several studies during this period have confirmed, once the relationship between GDP growth and energy consumption is corrected for energy quality, much of the historic evidence for energy-economy decoupling disappears. [19]

The Divisia index is a method for aggregating heat equivalents by their relative prices, and Figure 40 charts US GDP, Divisia-corrected energy consumption, and non-corrected energy consumption. According to Cutler Cleveland of the Center for Energy and Environmental Studies at Boston University, "This quality-corrected measure of energy use shows a much stronger connection with GDP [than non-corrected measures]. This visual observation is corroborated by econometric analysis that confirms a strong connection between energy use and GDP when energy quality is accounted for." [20]

Cleveland notes that, "Declines in the energy/GDP ratio are associated with the general shift from coal to oil, gas, and primary electricity." This holds true for many countries Cleveland and colleagues have examined. [21]

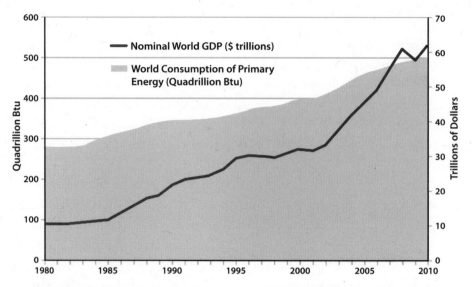

FIGURE 39. World Energy Use and GDP. Sources: US Energy Information Administration, International Monetary Fund.

His conclusion is highly relevant to our discussion of Peak Oil and energy substitution: "The manner in which these improvements have been... achieved should give pause for thought. If decoupling is largely illusory, any rise in the cost of producing high quality energy vectors could have important economic impacts.... If the substitution process cannot continue, further reductions in the E/GDP ratio would slow."[22]

Another often-ignored factor skewing the energy-GDP relationship is outsourcing of production. In the 1950s, the US was an industrial powerhouse, exporting manufactured products to the rest of the world. By the 1970s, Japan was becoming the world's leading manufacturer of a wide array of electronic consumer goods, and in the 1990s China became the source for an even larger basket of products, ranging from building materials to children's toys. By 2005, the US was importing a substantial proportion of its non-food consumer goods from China, running an average trade deficit with that country of close to $17 billion per month.[23] In effect, China was burning its coal to make America's consumer goods. The US derived domestic GDP growth from this commerce as Walmart sold mountains of cheap products to eager shoppers, while China expended most of the Btus. The American economy grew without using as much

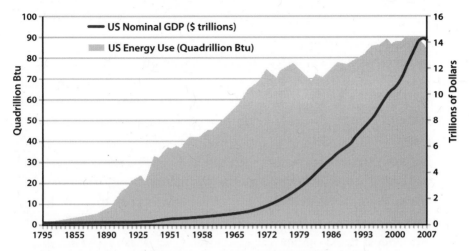

FIGURE 40A. Energy Use and Economic Growth in the United States, 1795–2009.
Sources: US Energy Information Agency, US Bureau of Economic Analysis, Louis
Johnston and Samuel H. Williamson, "What Was the US GDP Then?" MeasuringWorth,
2010.

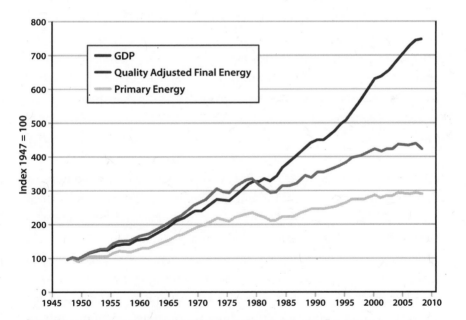

FIGURE 40B. Decoupling GDP and Energy Use in the US. Changes in energy quality
account for much of the divergence between energy consumption and GDP since
1980. Other factors include outsourcing of production and financialization of the
economy. Sources: US Energy Information Administration and Bureau of Economic
Analysis; Cutler Cleveland, Encyclopedia of Earth.

energy — *in America* — as it would have if those goods had been manufactured domestically.

There is one more factor that helps explain historic US "decoupling" of GDP growth from growth in energy consumption — the "financialization" of the economy (discussed in Chapter 2). Cutler Cleveland notes, "A dollar's worth of steel requires 93,000 BTU to produce in the United States; a dollar's worth of financial services uses 9,500 BTU."[24] As the US has concentrated less on manufacturing and building infrastructure, and more on lending and investing, GDP has increased with a minimum of energy consumption growth. While the statistics seem to show that we are becoming more energy efficient as a nation, to the degree that this efficiency is based on blowing credit bubbles it doesn't have much of a future. As we saw in Chapter 2, there are limits to debt.

The actual tightness of the relationship between energy use and GDP is illustrated in the recent research of Charles Hall and David Murphy at the State University of New York at Syracuse, which shows that, since 1970, high oil prices have been strongly correlated with recessions, and low oil prices with economic expansion. Recession tends to hit when oil prices reach an inflation-adjusted range of $80 to $85 a barrel, or when the aggregate cost of oil for the nation equals 5.5 percent of GDP.[25] If America had truly decoupled its energy use from GDP growth, then there wouldn't be such a strong correlation, and high energy prices would be a matter of little concern.

So far we've been considering a certain kind of energy efficiency — energy consumed per unit of GDP. But energy efficiency is more commonly thought of more narrowly as the efficiency by which energy is transformed into work. This kind of energy efficiency can be achieved in innumerable ways and instances throughout society, and it is almost invariably a very good thing. Sometimes, however, unrealistic claims are made for our potential to use energy efficiency to boost economic growth.

Since the 1970s, Amory Lovins of Rocky Mountain Institute has been advocating doing more with less and has demonstrated ingenious and inspiring ways to boost energy efficiency. His 1998 book *Factor Four* argued that the US could simultaneously double its total energy efficiency and

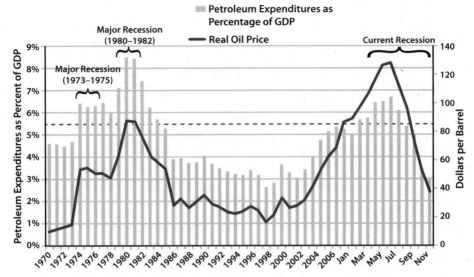

FIGURE 41A. From Hall and Murphy. The dotted line represents a threshold for petroleum expenditure as a percentage of GDP. When expenditures rise above this line, the economy begins to move towards recession. Distillate fuel oil, motor gasoline, LPG, and jet fuel are all included in petroleum expenditures. Source: Adapted from David J. Murphy and Charles A. S. Hall, 2011. "Energy return on investment, peak oil, and the end of economic growth" in *Ecological Economics Reviews*. Robert Costanza, Karin Limburg & Ida Kubiszewski, Eds. *Ann. N. Y. Acad. Sci.* 1219:52–72.

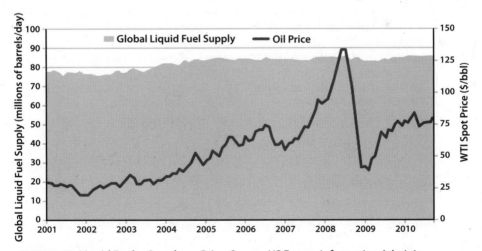

FIGURE 41B. Liquid Fuel — Supply vs. Price. Source: US Energy Information Administration.

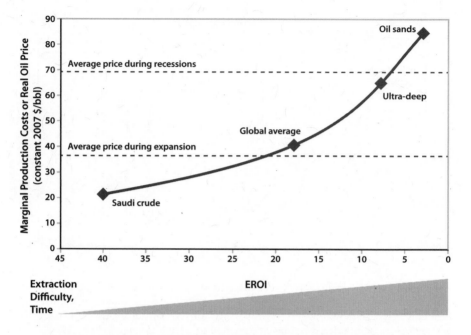

FIGURE 41C. Oil Production Costs from Various Sources as a Function of Energy Returned on Energy Invested (EROI). The dotted lines represent the real oil price averaged over both recessions and expansions during the period from 1970 through 2008. EROI data for oil sands come from Murphy and Hall, the EROI values for both Saudi Crude and ultradeep water were interpolated from other EROI data in Murphy and Hall, data on the EROI of average global oil production are from Gagnon et al., and the data on the cost of production come from Cambridge Energy Research Associates. Source: Adapted from David J. Murphy and Charles A. S. Hall. 2011. "Energy return on investment, peak oil, and the end of economic growth" in "Ecological Economics Reviews." Robert Costanza, Karin Limburg & Ida Kubiszewski, Eds. *Ann. N. Y. Acad. Sci.* 1219:52–72.

halve resource use.[26] More recently, he has upped the ante with "Factor 10" — the goal of maintaining current productivity while using only ten percent of the resources.[27]

Lovins has advocated a "negawatt revolution," arguing that utility customers don't want kilowatt-hours of electricity; they want energy services — and those services can often be provided in far more efficient ways than is currently done. In 1994, Lovins and his colleagues initiated the "Hypercar" project, with the goal of designing a sleek, carbon fiber-bodied hybrid that would achieve a three- to five-fold improvement in

fuel economy while delivering equal or better performance, safety, ame-
nity, and affordability as compared with conventional cars. Some innovations resulting from Hypercar research have made their way to market,
though today hybrid-engine cars still make up only a small share of vehicles sold.

While his contributions are laudable, Lovins has come under criticism
for certain of his forecasts regarding what efficiency would achieve. Some
of those include:

- Renewables will take huge swaths of the overall energy market (1976);
- Electricity consumption will fall (1984);
- Cellulosic ethanol will solve our oil import needs (repeatedly);
- Efficiency will lower energy consumption (repeatedly).[28]

The reality is that renewables have only nibbled at the overall energy
market; electricity consumption has grown; cellulosic ethanol is still in
the R&D phase (and faces enormous practical hurdles to becoming a primary energy source, as discussed above); and increased energy efficiency,
by itself, does not appear to lower consumption due to the well-studied
rebound effect, wherein efficiency tends to make energy cheaper so that
people can then afford to use more of it.[29]

Once again: energy efficiency is a worthy goal. When we exchange
old incandescent light bulbs for new LED lights that use a fraction of the
electricity and last far longer, we save energy and resources—and that's a
good thing. Full stop.

At the same time, it's important to have a realistic understanding of
efficiency's limits. Boosting energy efficiency requires investment, and
investments in energy efficiency eventually reach a point of diminishing
returns. Just as there are limits to resources, there are also limits to efficiency. Efficiency can save money and lead to the development of new
businesses and industries. But the potential for both savings and economic
development is finite.

Let's explore further the example of energy efficiency in lighting. The
transition from incandescent lighting to the use of compact fluorescents
is resulting in dramatic efficiency gains. A standard incandescent bulb
produces about 15 lumens per Watt, while a compact fluorescent (CFL)

can yield 75 l/w—a five-fold increase in efficiency. But how much more improvement is possible? LED lights currently under development should deliver about 150 l/w, twice the current efficiency of CFLs. But the theoretical maximum efficiency for producing white light from electricity is about 300 lumens per Watt, so only another doubling of efficiency is feasible once these new LEDs are in wide use.[30]

Moreover, energy efficiency is likely to look very different in a resource- and growth-constrained economy from how it does in a wealthy, growing, and resource-rich economy.

Permit me to use the example of my personal experience to illustrate the point. Over the past decade, my wife Janet and I have installed photovoltaic solar panels on our suburban house, as well as a solar hot water system. In the warmer months we often use solar cookers and a solar food dryer. We insulated our house as thoroughly as was practical, given the thickness of existing exterior walls, and replaced all our windows. We built a solar greenhouse onto the south side of the house to help collect heat. And we also bought a more fuel-efficient small car, as well as bicycles and an electric scooter.

All of this took years and lots of work and money (several tens of thousands of dollars). Fortunately, during this time we had steady incomes that enabled us to afford these energy-saving measures. But it's fair to say that we have yet to save nearly enough energy to justify our expenditures from a dollars-and-cents point of view. Do I regret any of it? No. As energy prices rise, we'll benefit increasingly from having invested in these highly efficient support systems.

But suppose we were just starting the project today. And let's also assume that, like millions of Americans, we were finding our household income declining now rather than growing. Rather than buying that new fuel-efficient car, we might opt for a 10-year-old Toyota Corolla or Honda Civic. If we could afford the PV and hot water solar systems at all, we would have to settle for scaled-back versions. For the most part, we would economize on energy just by cutting corners and doing without.

The way Janet and I pursued our quest for energy efficiency helped America's GDP: We put people to work and boosted the profits of several contractors and manufacturing companies. The way people in hard times will pursue energy efficiency will do much less to boost growth—and might actually do the opposite.

What was true for Janet and me is in many ways also true for society as a whole. America could improve its transportation energy efficiency dramatically by investing in a robust electrified rail system connecting every city in the nation.[31] Doing this would cost roughly $600 billion, but it would lead to dramatic reductions in oil consumption, thus lowering the US trade deficit and saving enormous amounts in fuel bills. At the same time, the US could rebuild its food system from the ground up, localizing production and eliminating fossil fuel inputs wherever possible. Doing so would increase the resilience of the system, the health of consumers, and the quality of the environment, while generating millions of employment or business opportunities.[32] The cost of this food system transition is difficult to calculate accurately, but it would no doubt be substantial.

However, the nation would be getting a late start on these efforts. We could fairly easily have afforded to do these things over the past few decades when the economy was growing. But building highways and industrializing and centralizing our food system generated profits for powerful interests, and the vulnerabilities we were creating by relying increasingly on freeways and big agribusiness were only obvious to those who were actually paying attention—a small and easily overlooked demographic category in America these days. As oil becomes more scarce and expensive, more of the real costs of our reliance on cars and industrial food will become apparent, but our ability to opt for rails and local organic food systems will be constrained by lack of investment capital. We will be forced to adapt in whatever ways we can afford. Where prior investments *have* been made in efficient transport infrastructure and resilient food systems, people will be better off as a result.

To the degree that energy efficiency helps us *adapt* to a shrinking economy and more expensive energy, it will be essential to our survival and well being. The sooner we invest in efficient ways of meeting our basic needs the better, even if it entails short-term sacrifice. However, to hope that efficiency will produce a continuous reduction in energy consumption while simultaneously yielding continuous economic growth is unrealistic.

Business Development: The Cavalry's on the Way

A remarkable book appeared in 2004 to almost no fanfare and little critical notice. The author was Mats Larsson, a Swedish business consultant, and his book was titled *The Limits of Business Development and Economic Growth*.[33] Unlike the thousands of business books published each year that promise to help managers become more effective, or that hint at new opportunities for profit, Larsson's conveyed a sobering message — one that the business community evidently didn't want to hear: Our human ability to invent genuinely new activities is probably limited, and most recent inventions have consisted merely of finding ways to speed up activities that humans have been performing for a very long time — communicating, transporting themselves and their goods, trading, and manufacturing. These processes can only be taken to the limits where things can be done at almost no time and at a very low cost, and we are fast approaching those limits.

"Through centuries and millennia," Larsson writes, "humans have struggled to simplify production and make tools and products less expensive and easier to manufacture." Possible examples are legion from virtually every industry — from telecommunications to air travel. "Now we are finally in a situation where many things can be done in close to no time and at a very low cost."[34] He goes on:

> [A]t close scrutiny we do not seem to have done anything except gradually automate activities that human beings have been performing for a few hundred, and sometimes thousand, years already. The development of a large number of different technologies that help us to automate these tasks has driven economic development and business proliferation in the past. Now, technological progress is at the stage where a number of these technologies and products have been developed to a point where we cannot realistically expect them to develop much further. And, despite widespread belief of the opposite, we cannot be certain that there are enough new products or technologies left to be developed for companies to be able to make use of the resources that are going to be freed from existing industries.[35]

For the skeptical reader such sweeping statements bring to mind the reputed pronouncement by IBM former president Tom Watson in 1943, "I think there is a world market for maybe five computers." Fortunes continue to be made from new products and business ideas like the iPad, Facebook, 3D television, BluRay DVD, cloud computing, biotech, and nanotech; soon we'll have computer-controlled 3D printing. However, Larsson would argue that these are in most cases essentially extensions of existing products and processes. He explicitly cautions that he is not saying that further improvements in technology and business are no longer possible — rather that, taken together, they will tend to yield diminishing returns for the economy as a whole as compared to innovations and improvements years or decades ago.

Fundamentally new technologies, products, and trends in business (as opposed to minor tweaks in existing ones) tend to develop at a slow pace. "Many of the big, resource-consuming trends of the near past are soon coming to an end in terms of their ability to attract investment and cover the cost of resources for development, production, and implementation."

Back in the late 1990s business was buzzing with talk of a "new economy" based on e-commerce. Internet start-up companies attracted enormous amounts of investment capital and experienced rapid growth. But while e-commerce flourished, many expectations about profit opportunities and rates of growth proved unrealistic.

Automation has reached the point where most businesses need dramatically fewer employees. "Presumably, this should make companies more profitable and increase their willingness to invest in new products and services," writes Larsson. "It does not. Instead, there is competition between more and more equal competitors, and all are forced to reduce prices to get their goods sold. The advantages of leading companies are getting smaller and smaller and it is becoming ever more difficult to find areas where unique advantages can be developed."[36] Regardless of the industry, "Most companies tend to use the same standard systems and more and more companies arrive at a situation where time and cost have been reduced to a minimum."[37]

It is bitterly ironic that so much success could lead to an ultimate failure to find further paths toward innovation and earnings.

Larsson estimated in 2004 that impediments to business development would begin to appear in the decade 2005–2015. His analysis did not take into account limits to the world's supplies of fossil fuels, nor declines worldwide in the amount of energy returned on efforts spent in obtaining energy. Neither did he examine limits to debt or declining ore quality for minerals essential to industry.[38] Remarkably, though, his forecast—based entirely on trends within business—points to an expiration date for global growth that coincides with forecasts based on credit and resource limits. This convergence of trends may not be merely coincidental: after all, automation is fed by cheap energy, and business growth by debt. Limits in one area tighten further the restrictions in others.

BOX 4.2   **How Much More Improvement Is Possible—Or Needed?**

A couple of centuries ago, if two people in distant cities wished to communicate, one of them traveled by foot, horse cart, or boat (or sent a messenger). It may have taken weeks. Now, using a cell phone, we can communicate almost instantly. A scholar might have had to travel to a distant library to access a particular piece of information. Today we can access all kinds of information via the Internet at almost no cost in almost no time. An ancient Sumerian would have used a clay tablet and wooden stylus for accounting. Now we use computers with business software to keep track of vastly more numerous transactions automatically in almost no time and at almost no cost.

While we are never likely to reach zero in terms of time and cost, we can be certain that *the closer we get to zero time and cost, the higher the cost of the next improvement and the lower the value of the next improvement will be.* This means that, with regard to each basic human technologically mediated pursuit (communication, transportation, accounting, and so on) we will sooner or later reach a point where the cost of the next improvement will be higher than its value.

We may be able to further improve the functionality of the Microsoft Office software package, the speed of transactions on the computer, computer storage capacity, or the number of sites available on the Internet. Yet on many of these development trajectories we will face a point when the value of yet another improvement will be lower than its cost to the consumer. At this point, further product "improvements" will be driven almost solely by aesthetic considerations identified by advertisers and marketers rather than by improvements achieved by engineers or inventors. For many consumer products this stage was reached decades ago.[39]

Moore's or Murphy's Law?

It is a truism in most people's minds that the most important driver of economic growth is new technology. Important innovations, from the railroad and the telegraph up through the satellite and the cell phone, have generated fortunes while creating markets and jobs. It may seem downright cynical to suggest that we won't see more of the same, leading to an abundant, technotopian future in which humanity has colonized space and all our needs are taken care of by obedient robots. But once again, there may be limits.

The idea that technology will continue to improve dramatically is often supported by reference to Moore's law. Over the past three decades, the number of transistors that can be placed inexpensively on an integrated circuit has doubled approximately every two years. Computer processing speed, memory capacity, and the number of pixels per dollar in digital cameras have all followed the same trajectory. This "law" is actually better thought of as a trend—but it is a trend that has continued for over a generation and is not expected to stop until 2015 or later. According to technology boosters, if the same innovative acumen that has led to Moore's law were applied to solving our energy, water, climate, and food problems, those problems would disappear in short order.

My first computer was an early laptop, circa 1986. It cost $1600 and had no hard disk—just two floppy drives—plus a small non-backlit, black-and-white LCD screen. It boasted 640K RAM internal memory and a processing speed of 9.54 MH. I thought it was wonderful! Its capabilities dwarfed those of the Moon-landing Apollo 11 spacecraft on-board computer, developed by NASA over a decade earlier.

My most recent laptop cost $1200 (that's a lot to pay these days, but it's a deluxe model), has a 250 gigabyte hard drive (holding about 200,000 times as much data as you could cram onto an old 3.5 inch floppy disk), 4 Gigs of internal memory, and a processing speed of 2.4 gigahertz. Its color LCD screen is stunning, and it does all sorts of things I could never have dreamed of doing with my first computer: it has a built-in camera so I can take still or moving pictures, it has sound, it plays movies — and, of course, it connects to the Internet! Many of those features are now standard even on machines selling for $300.

From 1986 to today, in just 25 years, the typical consumer-grade personal computer has increased in performance thousands of times over while dropping in price — noticeably so if inflation is factored in.[40]

So why hasn't the same thing happened with energy, transportation, and food production during this period? If it had, by now a new car would cost $750 and get 2000 miles to the gallon. But of course that's not the case. Is the problem simply that engineers in non-computer industries are lazy?

Of course not. It's because microprocessors are a special case. Moving electrons takes a lot less energy than moving tons of steel or grain. Making a two-ton automobile requires a heap of resources, no matter how you arrange and rearrange them. In many of the technologies that are critically important in our lives, recent decades have seen only minor improvements — and many or most of those have come about through the application of computer technology.[41]

Take the field of ground transportation (the example I'm about to use is also relevant to the energy efficiency and substitution discussions earlier in this chapter). We could make getting to and from stores and offices far more efficient by installing personal rapid transit (PRT) systems in every city in the world. PRT consists of small, automated vehicles operating on a network of specially built guide-ways (a pilot system has been built at Heathrow airport in London). The energy-per-passenger-mile efficiency of PRT promises to be much greater than that for personal automobiles, even electric ones, and greater even than for trolleys, streetcars, buses, subways, and other widely deployed forms of public transit. According to some estimates, a PRT system should attain an energy efficiency of 839 Btu per passenger mile (0.55 MJ per passenger km), as compared to the

3,496 Btu per passenger mile average for automobiles and 4,329 Btu for personal trucks.[42]

By the time we have shifted all local human transport to PRT, we may be approaching the limits of what is possible to achieve in terms of motorized, relatively high-speed transport energy efficiency. But to do this we will need massive investment, policy support, and the development of consumer demand. PRT may be an excellent idea, but its implementation is moving at a glacial pace—there's nothing "rapid" about it.

Far from already having implemented the most efficient transit systems imaginable, we find ourselves today even more dependent on cars and trucks than we were a half-century ago. Moreover, the typical automobile of 2011 is essentially similar to one from 1960: both are mostly made from steel, glass, aluminum, and rubber; both run on gasoline; both have similar basic parts (engine, transmission, gas tank, wheels, seats, body panels, etc.). Granted, today's car is more energy-efficient and sophisticated—largely because of the incorporation of computerized controls over its various systems. Much the same could be said for modern aircraft, as well as for the electricity grid system, water treatment and delivery systems, farming operations, and heating and cooling systems. Each of these is essentially a computer-assisted, somewhat more efficient version of what was already common two generations ago.

True, the field of home entertainment has seen some amazing technical advances over the past five decades—digital audio and video; the use of lasers to read from and record on CDs and DVDs; flat-screen, HD, and now 3D television; and the move from physical recorded media to distribution of MP3 and other digital recording formats over the Internet. Yet when it comes to how we get our food, water, and power, and how we transport ourselves and our goods, relatively little has changed in any truly fundamental way.

The nearly miraculous developments in semiconductor technologies that have revolutionized computing, communications, and home entertainment during the past few decades have led us to think we're making much more "progress" than we really are, and that more potential for development in some fields exists than really does. The slowest-moving areas of technology are, understandably, the ones that involve massive infrastructure that is expensive to build and replace. But these are the technologies on which the functioning of our civilization depends.

In fact, rather than showing evidence of great technological advance, our basic energy, water, and transport infrastructure shows signs of senescence, and of vulnerability to Murphy's law — the maxim that anything that can go wrong, will go wrong. In city after city, water and sewer pipes are aging and need replacement. The same is true of our electricity grids, natural gas pipes, roads, bridges, dams, airport runways, and railroads.

I live in Sonoma County, California, where officials declared last year that 90 percent of county roads will be allowed to deteriorate and gradually return to gravel, simply because there's no money in the budget to pay for continued repairs. Perhaps someone who lives on one of these Sonoma County roads will mail-order the latest MacBook Air (a shining aluminum-clad example of Moore's law) for delivery by UPS — only to be disappointed by the long wait because a delivery truck has broken its axle in a pothole (a dusty example of Murphy's law).

According to Ken Kirk, executive director of the National Association of Clean Water Agencies, more than 1,000 aging water and sewer systems around the US need urgent upgrades.[43] "Urgent" in this instance means that if infrastructure projects aren't undertaken now, the ability of many cities to supply drinking water in the years ahead will be threatened. The cost of renovating all these systems is likely to amount to between $500 billion and $1 trillion.

The failure of innovation and new investment to keep up with the decay of existing infrastructure is exemplified also in the fact that the world's global positioning system (GPS) is headed for disaster. Last year, the US Government Accountability Office (GAO) published a report noting that GPS satellites are wearing down and, if no new investments are made, the accuracy of the positioning system will gradually decline.[44] At some point during the next few decades, the whole system may crash. The GPS system happens to be one of the glowing highlights of recent technological progress. We depend on it not just for piloting Lincoln Navigators across the suburbs, but for guiding tractors through giant cornfields; for mapping, construction, and surveying; for scientific research; for moving troops in battle; and for dispatching emergency response vehicles to their appointed emergencies. How could we have allowed such an important piece of infrastructure to become so vulnerable?

There is one more reason to be skeptical about the capability of technological innovation across a broad range of fields to maintain economic growth, and though I have saved it to the end it is by no means a minor point. As verified in the research of the late Professor Vernon W. Ruttan of the University of Minnesota in his book *Is War Necessary for Economic Growth?: Military Procurement and Technology Development*, many large-scale technological developments of the past century depended on government support during early stages of research and development (computers, satellites, the Internet) or build-out of infrastructure (highways, airports, and railroads).[45] Ruttan studied six important technologies (the American mass production system, the airplane, space exploration and satellites, computer technology, the Internet, and nuclear power) and found that strategic, large-scale, military-related investments across decades on the part of government significantly helped speed up their development. Ruttan concluded that nuclear technology could not have been developed at all in the absence of large-scale and long-term government investments.

If, in the years ahead, government remains hamstrung by overwhelming levels of debt and declining tax revenues, investment that might lead to major technological innovation and infrastructure build-out is likely to be highly constrained. Which is to say, *it probably won't happen* — absent a wartime mobilization of virtually the entire economy.

We're counting on Moore's law while setting the stage for Murphy's.

Specialization and Globalization: Genies at Our Command

Economic efficiency doesn't flow from energy efficiency alone; it can also be achieved by increasing specialization or by expanding the scope of trade so as to exploit cheaper resources or labor. Both of these strategies have deep and ancient roots in human history.[46]

Division of labor increases economic efficiency by optimizing the use of people's unique talents, proclivities, and skills. If all people had to grow or gather all of their own food and fuel, the effort might require most of their working hours. By leaving food production to skilled farmers, we enable others to spend their days weaving cloth, playing the oboe, or screening hand-carried luggage at airports.[47] Prior to the agricultural revolution several millennia ago, division of labor was mostly along gender lines, and was otherwise part-time and informal; with farming and the settling of the first towns and cities, full-time division of labor appeared, along with social classes. Since the Industrial Revolution, the number of full-time occupations has soared.

If economists often underestimate the contribution of energy to economic growth, it would be just as wrong to disregard the role of specialization. Adam Smith, who was writing when Britain was still burning relatively trivial amounts of coal, believed that economic expansion would come about entirely because of division of labor. His paradigm of progress was the pin-making factory:

> I have seen a small manufactory of this kind where ten men only were employed.... But though they were very poor, and therefore but indifferently accommodated with the necessary machinery, they could, when they exerted themselves, make among them about twelve pounds of pins in a day. There are in a pound upwards of four thousand pins of a middling size. Those ten persons, therefore, could make among them upwards of forty-eight thousand pins in a day. Each person, therefore, making a tenth part of forty-eight thousand pins, might be considered as making four thousand eight hundred pins in a day. But if they had all wrought separately and independently, and without any of them having been educated to this peculiar business, they certainly could not each of them have made twenty, perhaps not one pin in a day....[48]

Later in *The Wealth of Nations*, Smith criticizes the division of labor, saying it leads to a "mental mutilation" in workers as they become ignorant and insular—so it's hard to know whether he thought the trend toward specialization was good or just inevitable. It's important to note, however, that it was under way *before* the fossil fuel revolution, and was already contributing to economic growth.

Standard economic theory tells us that trade is good. If one town has apple orchards but no wheat fields, while another town has wheat but no apples, trade can make everyone's diet more interesting. Enlarging the scope of trade can also reduce costs, if resources or products are scarce and expensive in one place but abundant and cheap elsewhere, or if people in one place are willing to accept less payment for their work than in another.

As mentioned in Chapter 1, trade has a long history and a somewhat controversial one, given that empires typically used military power to enforce trade rules that kept peripheral societies in a condition of relative poverty and dependency. The worldwide colonial efforts of European powers from the late 15th century through the mid-20th century exemplify this pattern; since then, enlargement of the scope of trade has assumed a somewhat different character and is now referred to as *globalization*.

Long-distance trade expanded dramatically from the 1980s onward as a result of the widespread use of cargo container ships, the development of satellite communications, and the application of computer technology. New international agreements and institutions (WTO, NAFTA, CAFTA, etc.) also helped speed up and broaden trade, maximizing efficiencies at every stage.

It may be that people were happier before trade, when they were embedded in more of a gift economy. But the material affluence of the world's wealthy nations simply cannot be sustained without trade at current levels. Indeed, without expanding trade, the world economy cannot grow. Period.[49]

Most economists regard division of labor and globalization as strategies that can continue to be expanded far into the future. To think otherwise would be to question the possibility of endless economic growth. But there are reasons to question this belief.

There may not be much more to achieve through specialization in the already-industrialized countries, as most tasks that can possibly be professionalized, commercialized, segmented, and apportioned already have been. Moreover, in a world of declining energy availability, the trend toward specialization could begin to be reversed. Specialization goes hand-in-hand with urbanization, and in recent decades urbanization has depended on surplus agricultural production from the industrialization of agriculture, which in turn depends on cheap oil.[50]

Relegating all food production to full-time farmers leaves more time for others to do different kinds of work. But in non-industrial agricultural societies, the class of farmers includes most of society. The specialist classes (soldiers, priests, merchants, scribes, and managers) are relatively small. It was only with the application of fossil fuels (and other strategies of intensification) to agriculture that we achieved a situation where, as in the US today, a mere two percent of the population grows nearly all the domestically produced food, freeing the other 98 percent to work at a dizzying variety of other jobs.

In other words, most of the specialization that has occurred since the beginning of the Industrial Revolution depended upon the availability of cheap energy. With cheap energy, it makes sense to replace human muscle-powered labor with the "labor" of fuel-fed machines, and it is possible to invent an enormous number of different kinds of machines to do different tasks. Tending and operating those machines requires specialized skills, so more mechanization tends to lead to more specialization.

But take away cheap energy and it becomes more cost-effective to do a growing number of tasks locally and with muscle power once again. As energy gets increasingly expensive, a countertrend is therefore likely to emerge: *generalization*. Like our ancestors of a century ago or more, most of us will need the kinds of knowledge and skill that can be adapted to a wide range of practical tasks.

Globalization will suffer a similar fate, as it is vulnerable not only to high fuel prices, but grid breakdowns, political instability, credit and currency problems, and the loss of satellite communications.

Jeff Rubin, the former chief economist at Canadian Imperial Bank of Commerce World Market, is the author of *Why Your World Is About To Get a Whole Lot Smaller: Oil and the End of Globalization*, which argues that the amounts of food and other goods imported from abroad will inevitably shrink, while long-distance driving will become a luxury and international travel rare.[51] Given time, he says, we could develop advanced sailing ships with which to resume overseas trade. But even then moving goods across land will require energy, so if we have less energy that will almost certainly translate to less mobility.

The near future will be a time that, in its physical limits, may resemble the distant past. "The very same economic forces that gutted our manufacturing sector," says Rubin, "that paved over our farm land, when oil was cheap and abundant, and transport costs were incidental, those same economic forces will do the opposite in a world of triple digit oil prices. And that is not determined by government, and that is not determined by ideological preference, and that is not determined by our willingness or unwillingness to reduce our carbon trail. That is just Economics 100. Triple digit oil price is going to change cost-curves. And when it changes cost-curves, it is going to change economic geography at the same time."[52]

Despite their economic advantages, specialization and globalization in some ways reduce resilience — a quality that is essential to our adapting to the end of growth. Extremely specialized workers may have difficulty accommodating themselves to the economic necessities of the post-growth world. In World War II, auto assembly plants in the US could be quickly re-purposed to produce tanks and planes for the war effort; today, when we need auto factories to make electric railroad locomotives and freight/passenger cars, the transition will be much more difficult because machines as well as workers are much more narrowly specialized. Moreover, dependence on global systems of trade and transport will leave many communities vulnerable if needed tools, products, materials, and spare parts are no longer available or are increasingly expensive due to rising transport costs. Successful adaptation will require economic re-localization and a generalist attitude toward problem solving.

Whether we like or hate globalization and specialization, we will nevertheless have to bend to the needs of an energy-constrained economy — and that will mean relying more on local resources and production capacity, and being able to do a broader range of tasks.

Of course, this is not to say that all activities will be localized, that trade will disappear, or that there will be no specialization. The point is simply that the recent extremes achieved in the trends toward specialization and globalization cannot be sustained and will be reversed. How far we will go toward being local generalists depends on how we handle the energy transition of the 21st century — or, in other words, how much of technological civilization we can preserve and adapt.

The near-religious belief that economic growth depends not on energy and resources, but solely on increasing innovation, efficiency, trade, and division of labor, can sometimes lead economists to say silly things.

Some of the silliest and most extreme statements along these lines are to be found in the writings of the late Julian Simon, a longtime business professor at the University of Illinois at Urbana-Champaign and Senior Fellow at the Cato Institute. In his 1981 book *The Ultimate Resource*, Simon declared that natural resources are effectively infinite and that the process of resource substitution can go on forever. There can never be overpopulation, he declared, because having more people just means having more problem-solvers.

How can resources be infinite on a small planet such as ours? Easy, said Simon. Just as there are infinitely many points on a one-inch line segment, so too there are infinitely many lines of division separating copper from non-copper, or oil from non-oil, or coal from non-coal in the Earth. Therefore, we cannot reliably quantify how much copper, oil, coal, or neodymium or gold there really is in the world. If we can't measure how much we have of these materials, that means the amounts are not finite—thus they are infinite.[53]

It's a logical fallacy so blindingly obvious that you'd think not a single vaguely intelligent reader would have let him get away with it. Clearly, an infinite number of dividing lines between copper and non-copper is not the same as an infinite quantity of copper. While a few critics pointed this out (notably Herman Daly), Simon's book was widely praised nevertheless.[54] Why? Because Simon was saying something that many people wanted to believe.

Simon himself is gone, but his way of thinking is alive and well in the works of Bjorn Lomborg, author of the bestselling book *The Skeptical Environmentalist* and star of the recent documentary film "Cool It."[55] Lomborg insists that the free market is making the environment ever healthier, and will solve all our problems if we just stop scaring ourselves needlessly about running out of resources.

It's a convenient "truth"—a message that's appealing not only because it's optimistic, but because it confirms a widespread, implicit belief that technology is equivalent to magic and can do anything we wish it to. Economists often talk about the magic of exponential growth of compound interest; with financial magic we can finance new technological magic. But in the real world there are limits to both kinds of "magic." Modern industrial technology has certainly accomplished miracles, but we tend to ignore the fact that it is, for the most part, merely a clever set of means for using a temporary abundance of cheap fossil energy to speed up and economize things we had already been doing for a very long time.

Many readers will say it's absurd to assert that technology is subject to inherent limits. They may recall an urban legend according to which the head of the US Patent Office in 1899 said that the office should be closed because everything that could be invented already had been invented (there's no evidence he actually did say this, by the way).[56] Aren't claims about limits to substitution, efficiency, and business development similarly wrong-headed now?

Not necessarily. Humans have always had to face social as well as resource limits. While the long arc of progress has carried us from knives of stone to Predator drones, there have been many reversals along the way. Civilizations advance human knowledge and technical ability, but they also tend to generate levels of complexity they cannot support beyond a certain point. When that point is reached, civilizations decline or collapse.[57]

I am certainly not saying that we humans won't continue to invent more new kinds of tools and processes. We are a cunning breed, and invention is one of our species' most effective survival strategies. However, the kinds of inventions we came up with in the 19th and 20th centuries were suited to human needs and interests in a world where energy and materials were cheap and amounts available were quickly expanding. Inventions of the 21st century will be ones suited to a world of expensive, declining energy and materials.

5 SHRINKING PIE: COMPETITION AND RELATIVE GROWTH IN A FINITE WORLD

... [C]ommerce is but a means to an end, the diffusion of civilization and wealth. To allow commerce to proceed until the source of civilization is weakened and overturned is like killing the goose to get the golden egg. Is the immediate creation of material wealth to be our only object? Have we not hereditary possessions in our just laws, our free and nobly developed constitution, our rich literature and philosophy, incomparably above material wealth, and which we are beyond all things bound to maintain, improve, and hand down in safety? And do we accomplish this duty in encouraging a growth of industry which must prove unstable, and perhaps involve all things in its fall?

— William Stanley Jevons (economist, 1865)

Is the central assertion of this book—that world economic growth is over—already disproved? How else to explain China's continued exuberant expansion, or signs of recovery in the US in 2010?

As stated in the Introduction, I am asserting that *real, aggregate, averaged* growth is essentially finished, though we may still see an occasional quarter or year of GDP growth relative to the previous quarter or year, and will still see residual growth in some nations or regions. The point can be summarized in a single sentence, but it bears reiterating and unpacking because there are several kinds of relative growth, and the competitive pursuit of advantage within a global economy that, overall, is shrinking rather than growing will powerfully shape political, geopolitical, and social developments for the next few decades.

In this chapter we will explore the growth prospects of the Asian economies. We will also examine the dynamics of currency wars. And we will see how rich and poor countries, and demographic sectors within those countries, are likely to fare in post-growth economy, and how increasing competition for depleting resources may drive nations toward conflict.

The best place to start this survey of prospects for short-to-medium-term relative growth is with China, which not only exemplifies rapid residual economic growth, but also points the way to how currency and resource rivalries, as well as old/young, rich/poor, urban/rural divisions might play out as the global economy contracts.

The China Bubble

If one were looking for a single arguing point against the idea that world economic growth is ending, China would almost certainly be the best choice. New Chinese cities are springing up in mere months. A stop-motion video posted on the Internet last year showed a 15-story hotel being built in six days.[1] A new coal power plant opens, on average, every four days. Twenty million Chinese move from the countryside to cities each year. Because city dwellers contribute 20 times as much per capita to GDP, urbanization alone accounts for half or more of China's 10 percent annual GDP growth. China is building highways faster than any other nation, and its motorists are now buying around 13 million automobiles per year, versus 11 million annually in the US—which had been the world's largest market for cars since the days of the Model T. It's all happening blindingly fast. Indeed, in both its scale and speed, the expansion of the Chinese economy is unprecedented in world history.

But how long can this go on? Will China escape the economic fate of older industrial nations, or is it poised for its own encounter with growth limits? There are four reasons for thinking current trends cannot be sustained.

1. Resource Depletion and Resource Competition:
The Story of China's Coal

China's appetite for resources and raw materials is driving up worldwide prices of a wide range of commodities including oil, iron, copper, cotton, cement, and soybeans. But for the Chinese economy, perhaps the single most important resource is coal. Indeed, it may not be an oversimplification to say that the fate of China's economy rests on its ability to maintain growth in coal supplies.

China relies on coal for 80 percent of its electricity and 70 percent of its total energy; coal also supports China's steel industry, the world's largest. Altogether, China is one of the most coal-dependent nations in the world. In order to become the world's second-largest economy, it has had to more than double its coal consumption over the past decade, so that it is now using nearly half of all coal consumed globally, and over three times as much as is consumed in the next nation in line, the US (which prides itself on being "the Saudi Arabia of coal").

As China energy expert David Fridley and I argued in a recent op-ed in *Nature*, while China claims it has enough coal to fuel continued economic growth, that claim is questionable.[2]

The nation has recently updated its proven coal reserves to 187 billion metric tons, putting it second in line after the US in terms of supplies. That would be about 62 years' worth of coal at 2009 rates of consumption (over three billion tons per year). But this simple "lifetime" calculation is highly misleading.

Reserves lifetime figures are calculated on the basis of flat demand and lose meaning if demand grows over time. China's coal consumption is accelerating rapidly, so that the expected "62 years' worth" must be adjusted downward. Demand forecasts from China's Energy Research Institute would reduce the reserves lifetime to about 33 years; but if coal demand were to grow in step with projected Chinese economic growth, the reserves lifetime would drop to just 19 years.

Yet this still doesn't capture the situation. Production will peak and decline long before China's coal completely runs out. Further, as with oil production, coal mining proceeds on the basis of the "best-first" or "low-hanging fruit" principle, so we must assume that China is extracting its highest-quality, easiest-accessed coal now, leaving the lower-quality and more expensively mined coal for later. Unlike the US, China does not have vast deposits of surface-minable coal; over 90 percent of China's coal comes from underground mines up to 1,000 meters in depth, and those mines face increasing engineering challenges.

Hubbert analysis, which has been used to forecast oil production peaks, can also be applied to forecasting future coal supplies. In 2007, Chinese academics Tao and Li forecast that China's coal production will peak and start to decline perhaps as early as 2025.[3] Other forecasts are more pessimistic. A 2007 analysis by the Energy Watch Group of Germany forecast a peak of production in 2015 with a rapid production decline commencing in 2020.[4] And a 2010 study by Patzek and Croft forecast the peak of *world* coal production for this year (2011); they see China's coal peak also occurring essentially now.[5]

China has few options for reducing its reliance on coal, since the fuel is used in so many ways. In addition to powering the electricity and steel sectors, coal provides winter heat to hundreds of millions of northern Chinese; it is also used in the cement, non-ferrous metals, and chemicals industries. While China is rapidly expanding its supply of natural gas, to replace just the coal used for heating would double total gas consumption.

China is quickly developing alternative energy sources. But can these be brought on line fast enough to make a difference? Let's do some numbers. China aims to have 100 gigawatts (GW) of wind power capacity by 2020, and the nation's leaders plan to expand installed solar capacity to 20 GW during the same period. These are truly astonishing goals, and, if China even comes close to accomplishing them, it will become the world's renewable energy leader. But there is a problem. Total Chinese electricity generation capacity is 900 GW currently; with seven percent growth, that means the nation's electricity demand in 2020 will be something like 1800 GW. Wind and solar together would supply less than seven percent of that. The only thing likely to boost that percentage much would be a dramatic reduction in growth of energy demand to, say, two percent annually.

The situation with nuclear power is similar: China has 11 atomic power plants now and is in the process of building 20 more, with a target of 60

GW of generating capacity, or possibly more, by 2020. But this will supply only between three and five percent of total electricity demand, depending on energy demand growth rates. In late 2010, energy policy makers in Beijing evidently began to take notice of the looming electricity supply problem, and rumors circulated of new efforts to construct up to 245 new nuclear plants over the next two decades (the US has only 104 in total). If this new target is real, and if the Chinese succeed in achieving it, a large fraction of new electricity demand for the coming years could be met through sources other than coal—but China would still have an enormous (though more slowly growing) coal dependence to feed. Meanwhile, China's soaring demand for uranium would push up global prices for this energy mineral.[6]

In 2009 China was a substantial net importer of coal, having been a net exporter every year through 2008.[7] China could import more coal to enable further growth, but the biggest exporters of coal—Australia, Indonesia, and South Africa—have much smaller reserves and production rates. The entire seaborne trade in steam coal (mainly used by power plants) currently amounts to only 630 million tons per year, and China could absorb this much with only three years of continued growth in coal demand. That's not going to happen, though: Other nations need that export coal, too—including India, also major coal-based economy, and also a country needing to import increasing amounts of fuel.

The conclusion is unsettling but inescapable: China's reliance on coal cannot be significantly reduced as long as its demand for electrical power continues to grow at anything like current rates. And even if energy demand growth tapers off and alternative energy sources come on line quickly, the country's ability to supply enough coal domestically will still be challenged. This will drive up coal prices worldwide, while choking off economic growth at home. China's energy economy is unsustainable and will cease growing in the foreseeable future, impacting many other nations as it does so.

2. Export-Led Development Model

The fundamental economic model that China has depended on for the past couple of decades was borrowed from Japan, and consists of producing low-cost export goods to fund investment at home. Essentially the same model is being pursued by Thailand, Vietnam, Taiwan, Malaysia, Singapore, the Philippines, and Indonesia. The story of what happened to Japan as a result of following this strategy should be a cautionary tale for its neighbors, and for Beijing in particular.[8]

The post-war Japanese export boom resulted in spectacular growth for four decades, but the undervalued yen eventually caused a deflationary contraction of Japan's economy, which is smaller today than it was in the late 1990s. There are good reasons to think the same policies will achieve the same results in China. However, China is much larger than Japan, and the world economy is today far more fragile than it was in the late 1980s when the Japanese bubble burst, so the global consequences of a Chinese crash would be far greater.

After World War II, Japan kept its yen weak, making exports relatively cheap to foreign buyers. Japan also benefited from a high savings rate, which enabled massive investment in infrastructure and manufacturing capacity. The country's GDP ballooned by 600 percent from 1950 to 1970, pulling more people out of poverty more quickly than had ever been done anywhere previously.

Export- and investment-driven growth typically discourages consumption, as domestic prices are kept high and salaries low (to help fuel exports). In the case of Japan, yields on savings were suppressed so that available capital would flow to corporations and the government.

All of this resulted in a lopsided economy. In most modern market economies, consumption accounts for around 65 percent of GDP (in the US, the proportion is 70 percent), while investment in fixed assets such as infrastructure and manufacturing capacity makes up 15 percent. In 1970, Japan's domestic consumption contributed 48 percent of the economy and fixed investment 40 percent. As Ethan Devine put it in his article "The Japan Syndrome" in *Foreign Policy*, "In plain English, the Japanese were consuming relatively little while investing heavily in steel plants and skyscrapers, which didn't leave much for fish or tourism. Belatedly, Tokyo realized that a balanced economy must also have consumption and that coating the country with factories and infrastructure wouldn't do the trick."[9]

Throughout the 1970s and early 1980s, Japan gradually strengthened the yen so as to support the development of a consumer culture, and consumption rose to more than half of GDP. In 1985, Tokyo let its currency appreciate more rapidly. But the result was simply a spectacular inflation of real estate and stock prices. The bubble's collapse lasted more than a decade, with stock prices scraping bottom in 2003 (before plummeting even further after the commencement of the current global crisis in 2008). Export-oriented industries could not adapt to a domestically led economy because there was insufficient consumer demand. And so rapid growth turned to stagnation, which has persisted up to the present.

Japan still runs on exports, but now government spending is an essential prop for the economy. Twenty years of fiscal stimulus have done little more than stave off even more serious economic contraction, while government debt has grown to nearly 200 percent of GDP.[10]

Fast forward to China, 2011. Like Japan, China subsists largely on exports while investing heavily in infrastructure, paying for the latter with private savings that come from tamping down consumption. Beijing adopted the Japanese growth model in the 1990s, when its deregulation and opening up of the country's economy was widely praised. While these policies created tens of millions of jobs, as well as thousands of new roads and millions of new buildings, they have also generated imbalances reminiscent of Japan in the 1980s — except that in many ways China has gone even further out on a limb.

Devine recites the startling numbers:

China is far more dependent on exports and investment than Japan ever was, and the numbers are still moving in the wrong direction. Investment accounts for half of China's economy while consumption is only 36 percent of GDP — the lowest in the world, drastically lower than even other emerging economies such as India and Brazil. But as the Japan example illustrates, low consumption leads to high savings, and China's thrifty citizens, coupled with booming net exports, have bestowed upon the country the world's largest current account surplus, triple that of Japan's in 1985.[11]

China's legendary trade surpluses cause problems for its trading partners while stoking price inflation at home. And inflation, the usual result of an undervalued currency, is dangerous in a country where hundreds of millions of people still have trouble affording basic essentials.

To outsiders, China has looked like a shining example of what growth can accomplish, yet it has achieved its success by strangling personal consumption (which was the engine of growth in the US and Europe) and sidelining small-scale entrepreneurs in favor of state-owned businesses and selected multinational corporations. Only a small percentage of its population has shared in the bounty.

China's leaders are aware of the pitfalls of pursuing the Japanese development model, and have issued a comprehensive slate of reforms to foster consumption and curb excessive capital investment. But these efforts will only work if the US and the rest of the world return to a path of growing consumption. If not, China's choices may be limited. An export-driven economy can only succeed if others can afford to import.

3. Demographics: Old/Young, Rich/Poor, Urban/Rural

Beijing's one-child policy, introduced in 1979, was largely effective — though it had the abhorrent side effect of encouraging a disdain for female infants, a prejudice that has led to abortion, neglect, abandonment, and even infanticide. Applying mainly to urban couples of Han descent, the policy reduced population growth in the country of 1.3 billion by as much as 300 million people. This meant that by the 1980s and '90s, young workers had fewer dependents to support — and China's manufacturing boom drew strength from young people moving from country to city to work in factories. For the nation as a whole, having a few hundred million fewer mouths to feed has acted as a social safety valve so far, and will reduce misery in the decades ahead as world resources deplete and human carrying capacity disappears.

However, there is a demographic price to pay. Beginning in 2015, China will see a growing number of older citizens relying on a shrinking pool of young workers.

Most of the nation's factories are located in its coastal cities, of which some, like Shenzen, were built from scratch as industrial centers. Shenzen hosts the Foxconn Technology Group, an electronics manufacturer that makes components for Dell, Hewlett-Packard, and Apple; nearly all its workers are under 25.

China's older workers have largely been left behind in rural villages, or pushed from their urban homes into apartment blocks on cities' outskirts to make way for new apartments and office buildings occupied by younger urbanites and the companies hiring them. Age discrimination is a fact of life.

All of this will gradually change as China's work force ages. Within a generation, the average age of a Chinese worker will be higher than that of an American worker.[12] One of China's leaders' biggest fears, expressed repeatedly in public pronouncements, is that the nation will grow old before it grows rich (Japan, in contrast, got rich before it grew old).

To avoid this fate, China is trying to grow its economy as fast as possible now, while it still can.

One way it does this is to offer paltry pensions and poor-quality health-care to older citizens. This makes China an attractive place for foreign corporations to do business. In the US, healthcare costs for older workers are often double the costs for workers in their 20s, 30s, and 40s. By keeping its workforce young and denying them benefits, China's leaders keep costs down. American or European companies that move production to China or buy Chinese goods gain leverage to rewrite terms of employment with their older workers at home — or they can simply shut down domestic factories.

China's youthful labor force attracts foreign investment. But as the country's work force ages, its competitive advantage may evaporate. Moreover, the lack of adequate pensions and healthcare for Chinese workers will eventually result in worsening social stresses and strains.

It is the financial sacrifices of its people that have given China the opportunity to attract capital investment to its industries, and that generate subsequent profits that are then loaned back to the United States and other industrialized nations.

To understand the significance of those sacrifices, one must understand a little of the country's recent history. At the end of the Communist revolution in 1949, China was impoverished and war-ravaged; the overwhelming

majority of its people were rural peasants. Communist Party chairman
Mao Zedong set a goal of bringing prosperity to the populous, resource-
rich nation. A period of economic growth and infrastructure development
ensued, lasting until the mid-1960s. At this point, Mao appears to have
had second thoughts: concerned that further industrialization would create or deepen class divisions, he unleashed the Cultural Revolution, lasting
from 1966 to the mid-1970s, when industrial and agricultural output fell.
As Mao's health declined, a vicious power struggle ensued, leading to the
reforms of Deng Xiaoping. Economic growth became a higher priority
than ever before, and it followed in spectacular fashion from widespread
privatization and the application of market principles. "To get rich is glori-
ous," Communist officials now proclaimed.

During the 1950s, '60s, and '70s, the Chinese people had worked hard
and endured grinding poverty for the good of the nation. But in the 1990s
a small segment of the populace — mostly in the coastal cities — began to
enjoy a middle-class existence. Some Chinese were indeed becoming glo-
riously rich, while most remained mired in extreme poverty. The resulting
wealth disparity is only bearable as long as the middle class continues
to expand in numbers, offering the promise of economic opportunity to
hundreds of millions of destitute peasants in the rural interior.

China's central government has unleashed a firestorm of entrepreneurial, profit-driven economic activity that is both unsustainable and difficult to control. Meanwhile, as we have seen, the uncontrollably dynamic
economy is export-dependent and ill suited to meeting domestic needs.

China has encouraged rapid export-led, coal-fired economic growth,
perhaps as a way of putting off dealing with its internal political, demographic, and social problems. If that is indeed Beijing's strategy, it has
worked spectacularly well for a short while. But it is built on contradic-
tions and false hopes. Over the course of the current decade, the Chinese
demographic-economic strategy will likely begin to unravel. What happens next is anybody's guess.[13]

4. Oh No — Not Another Real Estate Bubble!

There is one more similarity between Japan and China that is worth men-
tioning. During the 1980s, real estate prices in Tokyo were jaw-dropping.

In the Ginza district in 1989, choice properties fetched over 100 million yen (approximately $1 million US dollars) per square meter, or $93,000 per square foot. Prices were only slightly lower in other major business districts in the city. By 2004, values of top properties in Tokyo's financial districts had plummeted by 99 percent, and residential homes were selling for less than a tenth their peak prices. Tens of trillions of dollars in value were wiped out with the combined collapse of the Tokyo stock and real estate markets during the intervening years.

Once again, China is following in Japan's footsteps. Massive real estate projects — houses, shopping malls, factories, and skyscrapers — have been proliferating in China for years, attracting both private and corporate buyers. As prices have soared, investors have turned into speculators, intent on buying brand-new properties with the intention of flipping them.

Building is being driven by artificially inflated demand — the very definition of a bubble. And this is resulting in oversupply. In city after city, acres of commercial space sit vacant. Indeed, whole cities intended for millions of inhabitants have been built in the Chinese interior and now stand all but empty.[14] Some might argue that the Chinese are investing in infrastructure now in anticipation of many millions more citizens moving into urban centers over the coming decades — however, this presupposes continuing rapid economic growth, which is exactly what is in question. If growth sputters, this infrastructure overbuild will be a dead weight on the Chinese economy.

Though Beijing initiated an effort to cool the real estate and stock markets in 2008, the global financial crisis forced officials to relent in favor of lavish stimulus spending on shovel-ready infrastructure projects. The Chinese funneled 4 trillion yuan (about $590 billion) into what in many cases turned out to be yet more empty new shopping malls, empty new cities, and empty new factories.

For Chinese citizens, investment in the stock market hardly makes sense, given dramatic episodes of turbulence in recent years. Instead, a condominium or a house is seen as the most sensible and profitable investment. But this results in a bidding up of prices to the point where, in major cities like Beijing and Shanghai, a condo can cost 20 times a worker's annual salary. A worker in Tokyo might expect to pay only eight times her annual wages for a similar property.

What are the chances of putting off a property price meltdown? According to a November, 2010 article by Wieland Wagner in the German magazine *Der Spiegel*,

> Cao Jianhai of the Chinese Academy of Social Sciences in Beijing likens the Chinese economy to "a volcano before an eruption." Nevertheless, he doesn't believe that the government of Hu Jintao, the Communist Party leader and president, and Prime Minister Wen will allow a crash to occur before its term in office ends in 2012—local governments are too dependent on the real estate boom. According to Cao, Beijing will go to "any expense" to pump money into the financial system and spur a renewed surge of rapid economic growth.[15]

Once Hu and Wen are gone, however, it will be up to their successors to deal with the fallout from a housing crash.[16]

Economic Growth's Last Stand?

China is no more able to sustain perpetual growth than any other nation. The only questions, really, are when its growth will stall, and by what pace and to what degree its economy will contract.

The property bubble is likely to be China's biggest short-term problem, and it could have knock-on effects on the nation's banking system. The bubble could start to deflate as soon as next year, or the year after. Beijing will do what it can to prop up growth and tamp down social strain, and this could buy another couple of years—though there is no guarantee that the effort will succeed.

Over the longer haul (the next 2–10 years), China's greatest vulnerabilities are in the areas of energy, demographics, and the environment (water, climate, and agriculture). By the period 2016 to 2020, problems in these areas will accumulate and become mutually exacerbating, and it will eventually be impossible for China's leaders to plug all the leaks in the dike.

Already, China's social structure is stressed, as can be seen from the many regional rebellions that take place each year (but that go mostly unreported in world media). This is the main reason the central government is ruthless with respect to press and Internet freedoms and other civil liberties.

Talk to a businessperson from China and you may hear how the continued expansion of the Chinese economy is inevitable and unstoppable. But peer beneath the surface and you will see roiling, boiling ferment.

We have discussed China at some length, not only because it has become the world's second-largest national economy and is the world's foremost energy user, but because it is emblematic. India, Thailand, Indonesia, Malaysia, and Vietnam are each pursuing somewhat different paths toward the same grail of rapid economic growth, but their strategies and vulnerabilities are sufficiently similar that an understanding of China's predicament provides useful context for gauging these other countries' prospects.

China is likely the site of world economic growth's last stand. This nation, together with the other Asian "tigers," comprises the main engine of expansion that remains after the faltering of the older, more established economies in North America and Europe. When China sputters, the quickening slide of the global economy will be clear and obvious to everyone.

BOX 5.1   Is China Planning for the End of Growth?

China's new national 2011–1015 economic plan — which is essentially also its green blueprint — was finalized by the People's Congress in March 2011. The plan focuses on quality of development rather than on quantity only, with the goal of making the nation's economy less carbon-intensive and more resource-efficient through top-down mandates and regional pilot projects.

Beijing's 2010 energy efficiency goal was stringently enforced so that the nation would reach its target to decrease energy consumption per unit of GDP by 20 percent compared to 2005. As 2010 was winding down, some Zhejiang provincial officials tried to make their final five-year plan energy efficiency goal by enforcing rolling blackouts, turning off power at various times for days on end.

The new five-year plan prioritizes investments in renewable energy, information and communications technologies, advanced transportation and materials, water supply and treatment technologies (including using plants for bioremediation), and air and water quality.

China's move to become more of a service economy is likely inspired partly by a dawning awareness of limited resources and limited consumer demand in the West. Resource and energy efficiency and the shift to renewable energy sources may be driven by looming coal limits and rising oil prices. If leaders in Beijing are planning for less growth, it may be because they are beginning to recognize that current growth rates cannot be sustained. The strategies they are beginning to put in place are sensible from both an economic and an environmental standpoint. The question is: Can China's leaders put the brakes on growth fast enough to avoid looming obstacles, yet gradually enough to maintain control?[17]

Currency Wars

Since the economic crisis began, stresses in trade between the US and China have led to unfriendly official comments on both sides regarding the other nation's currency. Some financial commentators suggest that "currency wars," which might also embroil the European Union and other nations, may be in the offing, and that these could eventually turn into trade wars or even military conflicts. The US dollar, as the world's reserve currency and as the national currency of the country leading the world into the post-growth era, appears to be central to these "money wars."

It takes a little history to understand what currency conflicts are about.[18] Prior to the 20th century, most national currencies either consisted of gold or were tied to gold; therefore the currency of one nation was fairly easily convertible to that of another. National monetary reserves consisted of gold, and balance of payments deficits were settled in gold. Limited supplies of gold kept public spending within fairly tight bounds. Inflation through the debasement of a currency resulted in the refusal of other nations to accept that currency in trade. Typically the financing of wars presented the only exigency strong enough to overcome disincentives to debase money.

World War I, a conflict that engulfed at least 17 nations, was the first occasion when several countries simultaneously abandoned a hard money policy. Britain took on long-term war loans while Germany issued short-term bonds. Deficit financing arguably prolonged the war, resulting in millions of needless casualties.

Though Germany had entered the war with a thriving economy, its short-term debt, compounded by the harsh post-war terms of the Versailles Treaty, resulted in economic ruin through hyperinflation, leading to the destruction of its middle class and to the rise of Hitler, setting the stage for World War II.

At the Conference of Genoa in 1922, a partial return to the gold standard came about as the central banks of the world's powerful nations were permitted to keep part of their reserves in currencies (including the US dollar) that were directly exchangeable by other governments for gold coins. However, under this new Gold Exchange Standard, citizens could not themselves redeem national banknotes for gold coins. Now dollars and pounds were effectively equivalent to gold for the currency issuer, but not for most currency holders. This was an inherently inflationary development from a monetarist point of view (in that it meant that money could be issued substantially beyond the amounts of gold on deposit); however, the world's growing energy supplies and manufacturing capacity required an increase in the money supply, so for most countries and in most years measurable rates of price inflation remained relatively low.[19]

As World War II neared its end, Japan and the European powers lay in ruins; the United States was relatively unscathed. At the Bretton Woods monetary conference of 1944 the Allied nations laid the groundwork for a postwar international economic system that included new institutions such as the International Monetary Fund (IMF) and the International Bank for Reconstruction and Development (IBRD), which today is part of the World Bank. The US would assume a dominant role in these institutions, and the (partially) gold-backed dollar became, in effect, the world's reserve currency. Throughout the next half-century and more, citizens and businesses in nations around the world — even in the Soviet Union — who wanted a hedge against instability in their own national currency would hoard US greenbacks.

In the early 1970s, as the US borrowed heavily to finance the Vietnam War, France insisted on trading its surplus dollars for gold; this had the effect of emptying out US gold reserves. President Nixon's only apparent option was to ditch what remained of the gold standard. From then on, the dollar would have no fixed definition, other than as "the official currency of the United States."[20]

After 1973, many currencies kept a fixed exchange rate with the dollar. As of 2008, there were at least 17 national currencies still pegged to the US currency, including Aruba's florin, Jordan's dinar, Bahrain's dinar, Lebanon's pound, Oman's rial, Qatar's rial, as well as the Saudi riyal, Emirati dirham, Maldivian rufiyaa, Venezuelan bolivar, Belize dollar, Bahamian dollar, Hong Kong dollar, Barbados dollar, Trinidad and Tobago dollar, and Eastern Caribbean dollar.

While the US dollar now had no gold backing, in effect it was being backed by the oil of several key Middle East petroleum exporting nations, which sold their crude only for US dollars (thus creating and maintaining a worldwide demand for greenbacks with which to pay for oil) and then deposited their enormous earnings in US banks, which in turn made dollar-denominated loans throughout the world — loans that had to be repaid (with interest) in dollars.[21]

Meanwhile exchange rates for most currencies (including those of the European countries) floated relative to one another and to the dollar. This provided an opening for the emergence of the foreign exchange (ForEx) currency market, which has grown to an astonishing four trillion dollars per day in turnover as of 2010.

In 1999, most members of the European Union opted into a common currency, the euro, that floated in value like the Japanese yen. One of the motives for this historic monetary unification was the desire for a stronger currency that would be more stable and competitive relative to the US dollar.

For decades, China has been one of the countries that kept its currency pegged to the dollar at a fixed rate. This enabled the country to keep its currency's value low, making Chinese exports cheap and attractive — especially to the United States.

However, for smaller countries, fixed exchange rates have meant vulnerability to currency attacks. If speculators decide to sell large amounts of a country's currency, that country can defend its currency's value only by holding a large cache of foreign reserves sufficient to keep its fixed exchange rate in place. This reserve requirement effectively ties the country's leaders' hands during the attack, preventing them from spending (for example, to prop up banks); if the pegged exchange rate is abandoned under such circumstances, the currency's value will plummet. Either way, the nation faces the risk of economic depression or collapse — as occurred in the cases of the recent Argentine and East Asian financial crises.

Altogether, the world's currencies could hardly even be said to comprise a coherent "system": harmony and functionality are maintained only at great cost (with most of that cost ending up as profits to currency traders and speculators). But as world economic growth shifts into reverse, stresses within the global community of currencies may become unbearable.

With its enormous levels of public and private debt and its continuing trade deficits, the US has something to gain from a lower-valued dollar. This would make its export goods more attractive to foreign buyers; meanwhile, by making imports more expensive, it would help encourage savings and investment in domestic production. It would also enable the country to pay back its government debt with currency of lower value, effectively wiping out part of that debt. Maintaining low interest rates helps reduce the dollar's value, and the United States has kept interest rates low since the start of the crisis. But the US doesn't want to announce to the world that it is seeking to trash the dollar, because this could reduce the dollar's viability as the world's reserve currency — a status that yields multiple advantages to America's economy, and one that is increasingly being challenged.[22]

Investment money tends to "chase yield," which has the effect of driving up the value of the currencies in countries where investment opportunities and higher yields are to be found — currently, the young, industrializing countries of Asia. China and the other industrializing nations are responding by doing everything they can to keep exchange rates for their currencies low relative to the dollar so as to maintain trade advantages and reduce the impacts of an influx of yield-seeking money.

China has led the way in the international competition to weaken national currencies, but Japan and the US are seeking to lower the value of the yen and the dollar, respectively. According to Bill Black, writing in *Business Insider* on December 13, 2010,

> The EU, taking its lead from Germany, has allowed the Euro to ap-
> preciate against many currencies. Germany's high-tech exports can
> survive a strong Euro, but Greece, Spain, and Portugal cannot ex-
> port successfully under a strong Euro and their already severe eco-
> nomic crises can become much worse. The Irish will have serious
> problems, and their export problems would have been crippling
> if they were not a corporate income tax haven. Italy's, particularly
> southern Italy's, ability to export successfully is dubious.[23]

If the US dollar tumbles, that hurts China and other countries with fixed exchange rates; they feel pressured to drop their peg or revalue their currencies higher. Countries whose currencies are pegged to the dollar have had to resort to currency interventions and a massive buildup of foreign reserves to stop their currencies from appreciating. This is inflationary for those countries, and is one reason for the housing and equities boom in Asia.[24] China's way of pushing back against a lowering of the dollar's value is its threat of ceasing to purchase US Treasury debt (which it has in fact partly done). If neither the United States nor the industrializing nations back down, the result could be a final refusal of the latter nations to continue funding deficits in the US.[25]

As the US dollar has weakened, it has done so only against those cur-rencies that are free-floating. This has meant that countries like Japan and Germany have had to endure upward pressures on the value of their currencies. German Finance Minister Wolfgang Schäuble, interviewed in November 2010, had harsh words for his American counterparts, noting that "The US lived on borrowed money for too long," and adding that

The Fed's decisions [to buy US Treasury debt] bring more uncertainty to the global economy. They make it more difficult to achieve a reasonable balance between industrialized and emerging economies, and they undermine the US's credibility when it comes to fiscal policy. It's inconsistent for the Americans to accuse the Chinese of manipulating exchange rates and then to artificially depress the dollar exchange rate by printing money.[26]

Meanwhile, also in November 2010, China and Russia ceased using the dollar in bilateral trade, with Russian Prime Minister Vladimir Putin declaring that his country might eventually adopt the euro. Even though Russia is not one of China's top trading partners and is unlikely to be welcomed into the eurozone anytime soon, its leaders' hostility to the dollar helps exacerbate discontent elsewhere. If China excludes dollar trades with other primary non-US trade partners there may be a reason for Washington to worry. For now, Beijing appears merely to be letting off steam with no serious intent of isolating the United States, or of causing its nearly $3 trillion in US foreign exchange reserves to lose a significant portion of their value.

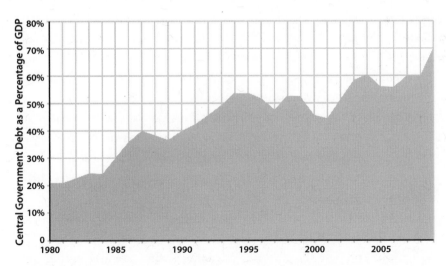

FIGURE 42. OECD Government Debt as a Percentage of GDP, 1980–2009. By 2009, the governments of the 34 OECD member countries held debts equal to 70 percent of their combined GDP. Source: Organization for Economic Co-operation and Development (OECD).

Thus for the time being, pundits who warn of wider and worse currency wars leading to trade or military conflicts may be exaggerating the threat.[27] Currency and trade wars are not in anyone's interest. A trade war between the US and China, for example, would reduce the GDP of both countries—and China would have more to lose than the United States. As long as cool heads prevail, currency conflicts are not likely to get out of hand. Of course, there is always the possibility that cooler heads may *not* prevail—especially in the politically volatile US, where members of Congress posture by threatening to refuse to raise the debt ceiling.

Over the longer term, the ecosystem of world currencies faces increasing dangers if growth fails to return in the US, and if the Chinese economic juggernaut falters.

Debt-based currencies that are traded without any clear international exchange standard create an inherently unstable situation. The so-called "goldbugs," economists who advocate a universal return to the gold standard, have plenty of grounds for criticizing free-floating currencies, but their alternative is simply not a realistic option: there isn't enough gold in the world to support anything like current levels of trade and investment, and much of the gold that exists is held in enormous reserves where it can do little good as a medium of exchange. The transition from the present system back to a gold standard would be intolerably chaotic, if it is even theoretically possible. Other kinds of fundamental national and global currency reforms (such as we will touch upon in the next chapter) may have better practical prospects over the long run, but are currently outside the realm of serious discussion among policy makers.

Without a return to economic growth, there is no sufficient remedy for the rapidly worsening stresses between and among the world's currencies. The lid can probably be kept on this boiling kettle in the short term, but over the course of the next decade it becomes more and more likely that something will give way.

Post-Growth Geopolitics

As nations compete for currency advantages, they are also eyeing the world's diminishing resources—fossil fuels, minerals, agricultural land, and water. Resource wars have been fought since the dawn of history, but today the competition is entering a new phase.

Nations need increasing amounts of energy and materials to produce economic growth, but—as we have seen—the costs of supplying new increments of energy and materials are increasing. In many cases all that remains are lower-quality resources that have high extraction costs. In some instances, securing access to these resources requires military expenditures as well. Meanwhile the struggle for the control of resources is re-aligning political power balances throughout the world.

The US, as the world's superpower, has the most to lose from a reshuffling of alliances and resource flows. The nation's leaders continue to play the game of geopolitics by 20th-century rules: They are still obsessed with the Carter Doctrine and focused on petroleum as the world's foremost resource prize (a situation largely necessitated by the country's continuing overwhelming dependence on oil imports, due in turn to a series of short-sighted political decisions stretching back at least to the 1970s). The ongoing war in Afghanistan exemplifies US inertia: Most experts agree that there is little to be gained from the conflict, but withdrawal of forces is politically unfeasible.

The United States maintains a globe-spanning network of over 800 military bases that formerly represented tokens of security to regimes throughout the world—but that now increasingly only provoke resentment among the locals. This enormous military machine requires a vast supply system originating with American weapons manufacturers that in turn depend on a prodigious and ever-expanding torrent of funds from the Treasury. Indeed, the nation's budget deficit largely stems from its trillion-dollar-per-year, first-priority commitment to continue growing its military-industrial complex.

Yet despite the country's gargantuan expenditures on high-tech weaponry, its armed forces appear to be stretched to their limits, fielding around 200,000 troops and even larger numbers of support personnel in Iraq and Afghanistan, where supply chains are both vulnerable and expensive to maintain.

In short, the United States remains an enormously powerful nation militarily, with thousands of nuclear weapons in addition to its unparalleled conventional forces, yet it suffers from declining strategic flexibility.

The European Union, traditionally allied with the US, is increasingly mapping its priorities independently—partly because of increased energy dependence on Russia, and partly because of economic rivalries and currency conflicts with America. Germany's economy is one of the few to have emerged from the 2008 crisis relatively unscathed, but the country is faced with the problem of having to bail out more and more of its neighbors. The ongoing European serial sovereign debt crisis could eventually undermine the German economy and throw into doubt the long-term soundness of the euro and the EU itself.[28]

The UK is a mere shadow of its former imperial self, with unsustainable levels of debt, declining military budgets, and falling oil production. Its foreign policy is still largely dictated in Washington, though many Britons are increasingly unhappy with this state of affairs.

China is the rising power of the 21st century, according to many geopolitical pundits, with a surging military and lots of cash with which to buy access to resources (oil, coal, minerals, and farmland) around the planet. Yet while it is building an imperial-class navy that could eventually threaten America's, Beijing suffers (as we have already seen) from domestic political and economic weaknesses that could make its turn at the center of the world stage a brief one.

Japan, with the world's third-largest national economy, is wary of China and increasingly uncertain of its protector, the US. The country is tentatively rebuilding its military so as to be able to defend its interests independently. Disputes with China over oil and gas deposits in the East China Sea are likely to worsen, as Japan has almost no domestic fossil fuel resources and needs secure access to supplies.

Russia is a resource powerhouse but is also politically corrupt and remains economically crippled. With a residual military force at the ready, it vies with China and the US for control of Caspian and Central Asian energy and mineral wealth through alliances with former Soviet states. It tends to strike tentative deals with China to counter American interests, but ultimately Beijing may be as much of a rival as Washington. Moscow uses its gas exports as a bargaining chip for influence in Europe. Meanwhile, little of the income from the country's resource riches benefits the populace. The Russian people's advantage in all this may be that they have recently been through one political-economic collapse and will therefore be relatively well-prepared to navigate another.

Even as countries like Venezuela, Bolivia, Ecuador, and Nicaragua reject American foreign policy, the US continues to exert enormous influence on resource-rich Latin America via North American-based corporations, which in some cases wield overwhelming influence over entire national economies. However, China is now actively contracting for access to energy and mineral resources throughout this region, which is resulting in a gradual shift in economic spheres of interest.

Africa is a site of fast-growing US investment in oil and other mineral extraction projects (as evidenced by the establishment in 2009 of Africom, a military strategic command center on par with Centcom, Eucom, Northcom, Pacom, and Southcom), but is also a target of Chinese and European resource acquisition efforts. Proxy conflicts there between and among these powers may intensify in the years ahead — in most instances, to the sad detriment of African peoples.[29]

The Middle East maintains vast oil wealth (though reserves have been substantially overestimated due to rivalries inside OPEC), but is characterized by extreme economic inequality, high population growth rates, political instability, and the need for importation of non-energy resources (including food and water). The revolutions and protests in Tunisia, Egypt, Libya, Bahrain, and Yemen in early 2011 were interpreted by many observers as indicating the inability of the common people in Middle Eastern regimes to tolerate sharply rising food, water, and energy prices in the context of autocratic political regimes.[30] As economic conditions worsen, many more nations — including ones outside the Middle East — could become destabilized; the ultimate consequences are unknowable at this point, but could well be enormous.

Like China, Saudi Arabia is buying farmland in Australia, New Zealand, and the US. Nations like Iraq and Iran need advanced technology with which to maintain an oil industry that is moving from easy plays to oilfields that are smaller, harder to access, and more expensive to produce, and both Chinese and US companies stand ready to supply it.

The deep oceans and the Arctic will be areas of growing resource interest, as long as the world's wealthier nations are still capable of mounting increasingly expensive efforts to compete for and extract strategic materials in these extreme environments.[31] However, both military maneuvering and engineering-mining efforts will see diminishing returns as costs rise and payoffs diminish.

Unfortunately, rising costs and flagging returns from resource conflicts will not guarantee world peace. History suggests that as nations become more desperate to maintain their relative positions of strength and advantage, they may lash out in ways that serve no rational purpose.

Again, no crisis is imminent as long as cool heads prevail. But the world system is losing stability. Current economic and geopolitical conditions would appear to support a forecast not for increasing economic growth, democracy, and peace, but for more political volatility, and for greater government military mobilization justified under the banner of security.

Population Stress: Old vs. Young on a Full Planet

Throughout the past two centuries economic growth has translated to an increased capability to support more humans with Earth's available resources. More energy, more raw materials, more jobs, more trade, better sanitation, and key medical advances have all contributed to higher infant survival rates and longer life expectancy in general. Human population growth can be seen as an indication of our success as a species.[32]

But now, as economic growth ends, higher population levels pose an enormous vulnerability. Declining energy, declining minerals and fresh water, and reduced global trade will challenge our ability to maintain existing food and public health systems, perhaps even in currently wealthy countries.

August Comte, the 19th-century French sociologist, famously declared that, "demography is destiny." During the coming post-growth decades, the nations of the world will face somewhat differing challenges depending on their size of population, rates of population growth, median age, and degree of urbanization.

Countries with large, youthful, and growing urban populations will be hardest hit. Young people will face lack of economic opportunity as trade contracts. Also, countries with young populations will see continuing population growth even if efforts are undertaken now to rein in fertility, simply because the bulk of the population will be in the child-bearing age range for the next two or three decades.

Countries with stable or declining populations (this includes most western European nations) will see aging populations, and thus a declining proportion of the population will consist of youthful workers (as we saw in the case of China). Some economists see this as a serious problem, and as a result Germany is offering cash financial incentives for couples to reproduce. However, this merely puts off inevitable process of adjustment to the end of population growth.

The end of economic growth will pose demographic challenges to all societies. But having more people will result in a bigger challenge than having fewer.

In a low-income society, when people have many children they tend to spend whatever money they have on keeping those children fed, so there is little left over to invest in future economic productivity (including education for children). This is a situation that tends to lead to continuing poverty. If there is no surplus income, there is nothing for the government to tax, so governments don't expand infrastructure: they don't build roads to rural areas so farmers can get their product to market — or water treatment facilities, or electricity grids, or schools. If farmers can't get their products to market, they may eventually give up and move to the cities where they strain whatever support infrastructure does exist. One of the best hopes for a society in this kind of bind is to reduce fertility.

Since World War II, eight countries (Tunisia, Japan, South Korea, Singapore, Taiwan, Barbados, Hong Kong, and Bahamas) have achieved the shift from being listed as "developing" to "developed" by first bringing fertility down through strong family planning programs. Once fewer children were being born, families found that they had money left over after paying for basic necessities, and this led to capital formation through personal savings. Demographers call this the "demographic dividend."

The continent of Africa will probably encounter the worst demographic challenges of any region in the decades ahead. Its population is expected to double its numbers by 2050, according to the UN. By then, Africa's urban population may have tripled, with 1.3 billion living in cities. These trends of rapid population growth and rapid urbanization cannot be sustained in a world of declining energy, scarce water, and changing climate, and will soon become enormous liabilities as today's quickly growing slums turn into centers of even greater human misery.

South Asia will also encounter enormous problems. Especially vulnerable is Pakistan, whose rapid population growth is already undermining access to education and medical facilities while posing serious health problems for women.[33]

The US has the fastest growing population of any industrialized country—mostly due to immigration (though immigration rates have declined in the last couple of years, probably due to the economic crisis). Already a hot-button issue, immigration could become even more of one as the economy contracts. But further waves of immigrants are possible if Mexico's economy fails due to declining revenues from oil production.

Further declines in the US economy will shift public opinion toward wanting to restrict immigration and population growth. Every survey since the 1940s has shown that a majority of Americans favors reducing immigration, yet during that time legal immigration has quadrupled (it doubled during Bush I and again during Bush II). Much of the support for liberalizing immigration policy has come from the Democratic party (in its calculus, more immigrants mean more Democrats), as well as from the construction industry (more immigrants equal more housing starts), the food industry (which depends on low-paid seasonal farm workers), and the US Chamber of Commerce (immigrants reduce labor costs).

Sadly, the debate has failed to take account of one key question: What is the population level the US can sustain? By most accounts, the country is already overdrawing resources, so that future generations will have restricted access to fresh water, fertile soil, and useful minerals. Adding more people through immigration simply steals further from our grandchildren. Gains in the efficiency with which resources are used may help temporarily, but population growth erases those gains over the long run.

For all nations, immigration laws need to be based on reasonable targets based in turn on estimates of human carrying capacity. Those laws must deal humanely with extreme circumstances, including provisions for refugees—such as climate refugees, whose numbers will likely multiply dramatically in the years ahead.

Meanwhile, declining economic growth will probably lead to increased demographic competition between the old (who will be seen by the young to have used up the world's resources) and the young (who will be seen by the old as a threat to savings and economic stability).

In the US, this competition may already be taking political form through the Tea Party movement, whose main agenda is to end government borrowing, bailouts, and stimulus packages, and to cap the national debt. These priorities are attractive to older, wealthier citizens who are concerned about protecting their savings from inflation—which would tend to benefit younger people saddled with debt.[34] Meanwhile, younger citizens are unhappy looking toward a future in which college and home ownership are no longer affordable and few jobs are available.

The population problem is solvable by making family planning and contraception freely available, changing cultural norms (as is being done by Population Media Center), and advancing women's rights.[35] But consider a best-case scenario: In a dozen years, given proper funding, virtually all countries could achieve a replacement level of fertility. Still, even after that monumental accomplishment, it would be another 70 years before the world as a whole would achieve zero population growth or begin a controlled decline.

The population issue has been highly politicized, and those who argue for controls on population growth are often demonized as elitist, racist, or misogynist. This is tragic, because the ongoing debate has caused humanity to put off dealing with the problem for far too long. And it is the poor, and especially poor women and children, who will pay the price for this delay.

BOX 5.2 | **Implications for Women**

During the past two centuries, industrialization and urbanization resulted in women's entry into the work force, and this in turn catalyzed the movement for women's rights. The end of cheap energy could see a return of women to work centered primarily in the home, and a consequent reduction in women's rights and opportunities, unless attention and effort are devoted by both men and women to averting that outcome.

In her 2004 essay, "Peak Oil Is a Women's Issue," Sharon Astyk wrote:

"Whatever happens in the post peak future will hit women differently, and in many ways harder, than it will hit men. For example, women are more likely to be poor than men are. In an economic crisis, women are more likely than men to be impoverished, and more seriously. Elderly women are the poorest and most vulnerable people in the US, and their lives are not likely to be improved by peak oil. Women are more likely to be single parents, a job that will come with a whole host of new difficulties post peak. They are more likely than men to work minimum wage jobs, to be exploited at work.... Poor women are more likely to be victims of violence, to have unplanned children, to be trapped in poverty from which they can't arise. In a period of economic crisis, where everyone is desperate for work, women will be even more vulnerable than usual, and we are already more vulnerable than men.

"Creating a sustainable future requires that women who don't want to have children, or not yet, or not many, be able to cease doing so. And yet poverty dramatically decreases access to medical care and birth control even in our first-world society. The poorer and less well-educated you are (and those two things are reciprocally related) the more likely you are to become pregnant without intending it, both because of reduced access to reliable birth control and insufficient education in how to use it. The younger, poorer and less well-educated a woman is, the younger she is likely to have children, the more children she is likely to have, the more health consequences she and her children are likely to have (prematurity, high blood pressure, etc...), and the less likely she is to ever escape poverty—or for her children to escape it. In a major economic depression, the ranks of poor women are likely to grow enormously, and we are likely to see not fewer children, but more and more unwanted children unless we plan very carefully to ensure that we prioritize medical access for everyone as one of the things we do with our limited resources."[36]

Astyk also points out that "Women also still do a disproportionate amount of child-rearing in just about every society, especially among very young children. They are the ones who instill values and ethics in their children to a large degree."

The population issue is discussed from a women's perspective in the new documentary film "Mother: Caring Our Way Out of the Population Dilemma."[37]

The End of "Development"?

For decades agencies that either actually or ostensibly aimed to aid impoverished nations have employed the terms "developed," "developing," and "underdeveloped" to refer to countries at various stages of industrialization. In ordinary usage, the word *develop* often means "to progress from an embryonic to an adult form"; thus its application to processes of economic and social change has conveyed an implicit assumption of inevitability. By calling rich industrialized countries "developed" and poor non-industrial countries "underdeveloped," policy makers were in effect saying that industrialization is equivalent to the healthy biological process of maturation, and should be the goal of all human societies. Through a trade-led process of economic expansion, non-industrial countries with subsistence economies and large indigenous populations must aim to become urbanized, consumer-driven, cosmopolitan manufacturing centers (according to this view): it is their right and destiny to do so.

This set of assumptions was always questionable. Indeed, it has been attacked with some vigor by Vandana Shiva, Helena Norberg-Hodge, Martin Kohr, Jerry Mander, Doug Tompkins, Gustavo Esteva, Edward Goldsmith, Ivan Illich, Manfred Max-Neef, David Graeber, and other prominent development critics (sometimes also known as post-development theorists).[38]

The critics of development claimed that the project of using loans and aid packages to fund huge infrastructure projects in poor nations, or to build factories there for multinational corporations, was at its core merely a continuation of colonialism by other means. Since two-thirds of the world's nations were defined as "underdeveloped," this meant that people in most countries needed to look outside of their own cultures for economic, agricultural, and educational models. Poor Third World nations were encouraged to take on enormous amounts of debt, and to flush young people out of the countryside and into cities. All of this came at both a human and an environmental cost.

Development, according to the critics, was actually a euphemism for post-war American hegemony; and indeed it was the US (along with its European allies) that provided the loans, trade rules, educational templates, and media images that would reshape societies across the global south.

BOX 5.3   **Development and Freedom**

Nobel Prize-winning economist Amartya Sen argues that a country's development should be measured by the freedom of its citizens.[42] Sen points to political freedom, civil rights, economic freedom, social opportunities (access to healthcare, education, and other social services), transparency guarantees (dealings with others and the government that are characterized by a mutual understanding of what is expected and what is offered), and protective security (unemployment benefits, famine and emergency relief, and general social safety nets) as comprising an interdependent bundle of freedoms that are instrumental in enabling people to live better lives.

Sen refers to these freedoms as "capabilities," arguing that they contribute to human functioning.[43] He sees life as a set of "doings and beings"—i.e., being healthy, being employed, being safe, and so on. Capabilities are a person's ability to do or be what they want given the resources they have. Civil rights, government transparency, education, and famine relief all make people better able to do and be what they want with what they have. And, for Sen, it is the ability and freedom to achieve what one wants that is the hallmark of development. Within the capabilities approach, development can be understood then as a process by which capabilities and freedom are expanded—not one where large increases are achieved in national GDP.

In this view, developed countries are not those with the highest per capita incomes, but those where people are best able to do and to be what they want.

One way to understand capabilities, development, and freedom is to think of them in the context of food security. It is widely understood that today people typically go hungry not from lack of food, but from their inability to access it. Poverty, social exclusion, and corrupt governance can all result in people being denied access to food. For example, a person may have the money to buy food and still go hungry due to unequal social status that results in exclusion from food networks. The way to remedy this is not by money alone, but by expanding capabilities and freedom. What the individual needs is civil rights and equal protection under the law. And the best way to achieve this, according to Sen, is through democracy. Sen argues that there has never been a famine under a democratic government. This is because in a democratic system people are better able to petition the government to act in emergency situations; also, democratic accountability provides incentives for leaders to act.

Sen's capabilities approach has been incorporated into some welfare and poverty measurements being used today. The best example is the UN Human Development Index (HDI). This measure looks at education and life expectancy along with income to measure and understand a country's development. It is interesting to note that the HDI does not correspond directly with a ranking of countries by GDP, or even GDP per capita—a measure that tends to mask inequality. For example, Norway ranks first on the HDI and 25th by national GDP. Conversely, China ranks second by national GDP and 89th on the HDI.

Another important consequence of the capabilities approach is that it allows for some real welfare improvements to be made that do not rely on increased resource use. For example, rooting out political corruption and ineptitude does not require much energy or resource throughput. However, other "capabilities," including education, health, and nutrition may be more problematic in this regard. In their paper "Energetic Limits to Economic Growth," Davidson, Brown, et al. argue that it has not been possible to increase socially desirable goods and services like nutrition, education, healthcare, technology, and innovation without increasing the consumption of energy and other natural resources.[44]

Two books galvanized anti-globalization activism and epitomized the arguments of development critics: *Ancient Futures* (1991) by Helena Norberg-Hodge, and *Confessions of an Economic Hit Man* (2005) by John Perkins.[39]

As a graduate student in linguistics in the 1970s, Norberg-Hodge had chosen to do field work in Ladakh, a remote Buddhist region in northern India. She found there a traditional village-based society virtually untouched by the modern Western world. The people had their problems, as people everywhere do, but the culture had evolved to suit its ecological constraints and opportunities; most people seemed generally happy, helpful, and friendly. Norberg-Hodge has continued her work in Ladakh up to the present, documenting how development has uprooted families, upended cultural norms, turned self-sufficiency into dependency, and created more misery than satisfaction.

John Perkins, a former chief economist at a Boston-based strategic consulting firm, claims he was groomed in the 1970s as an "economic hit man," in which capacity he helped needlessly to plunge poor nations like Indonesia and Panama into debt. Perkins tells how he used purposefully over-optimistic economic projections to persuade foreign governments to accept billions of dollars in loans from the World Bank and other institutions in order to build dams, airports, electric grids, and other infrastructure that he knew they couldn't afford and didn't need. Construction and engineering contracts were routed to US companies, with bribes to top foreign officials smoothing the way. However, the resulting debts ultimately had to be shouldered by the taxpayers in the poor countries. When payments couldn't be made, the World Bank or International Monetary Fund would jet in a team of economists to dictate the country's budget and security agreements. Perkins contends that this all amounted to a clever way for the US to expand its global influence at the expense of citizens in poor, often largely indigenous nations.

The defenders of development have always maintained that such claims are either fabricated or overblown, and that the real purpose of loans and aid packages has been to raise the standard of living of people in the world's poorest countries. Statistics lend support to this view. The

---

BOX 5.4	Development or Overdevelopment?

Environmental preservationist and publisher Doug Tompkins, among others, uses the term "overdeveloped" to refer to what are conventionally called "developed" nations, and "nations on the road to overdevelopment" for those usually classified as "developing." In this view, the goal of "development" as typically pursued is a condition that is clearly unsustainable, hence the term used to refer to it should reflect that fact.[45] The term also is logically required as a counterpart to the more commonly employed concept of "underdevelopment." Questioning how and why nations develop unevenly can lead to greater insight into the process whereby certain nations commandeer resources and labor to become "overdeveloped."

---

recent 2010 UN Human Development Report concludes that people in poor nations are generally healthier, wealthier, and better educated than they were 40 years ago.[40] Surveying human progress in 135 countries — 92 percent of the world's population — the report shows that average life expectancy rose from 59 to 70 years, primary school enrolment grew from 55 to 70 percent, and per capita income doubled to more than $10,000. The report's authors do devote a section to discussion of the "weak association between [GDP] growth and quality of life indicators such as health, education, political freedom, conflict and inequality," and also note that, "Within countries rising income inequality is the norm."

Development critics place significant emphasis on these latter points. It is relatively easy to measure GDP; it is more difficult to quantitatively assess the integrity of families and communities. Also, much of the measured "progress" of the past few decades has occurred in a few rapidly industrializing nations, while many very poor countries have actually lost ground in terms of most citizens' access to food, water, and shelter. Averages and totals can obscure important and worrisome details.

The discussion about development's efficacy was important during the heyday of economic expansion. However, now that growth is ending, the goal of conventional economic development—whether or not it ever made sense from humanitarian and environmental points of view—may have become largely unachievable.[41]

Without cheap transport fuel, the advantages of globalization begin to disappear, as we saw in Chapter 4. Scarce and expensive petroleum also undermines the project of industrializing agriculture and makes highway construction nonsensical. And the global credit crisis spells an end to big loans for unneeded infrastructure projects.

The priorities of poor nations have just changed. From here on, competing in the global economy will recede in importance; the primary challenge will be to adapt to post-growth world economic conditions and ever-worsening environmental challenges.

The UN Human Development Report does not address the resilience of societies in the face of declining global energy resources, the rising scale environmental disasters, or the end of economic growth (only the potential impacts of climate change are mentioned, though not taken fully into account). But in the decades ahead resilience will count for far more than economic competitiveness.

Urbanization, the industrialization of food systems, and the building of highways may have contributed to GDP over the short term, but they have created societal vulnerability over the longer term. In a world of Peak Oil, scarce fresh water, unstable currencies, changing climate, and declining trade, true "development" may require implementation of policies at odds with—sometimes the very reverse of—those of recent decades.

The phrase "sustainable development" entered the development lexicon in the late 1980s with the publication of the Report of the Brundtland Commission (the World Commission on Environment and Development); it was defined as development that "meets the needs of the present without compromising the ability of future generations to meet their own needs." While this ideal has often been watered down in the process of implementation, if taken seriously it could help poor nations identify sound strategies for dealing with the end of growth.

All the world's nations need to continue solving basic human problems (i.e., providing food, education, healthcare, and security) in the face of global environmental and economic change, without drawing down

Earth's nonrenewable resources or saddling future generations with oner-ous debt. And they have to do this in a way that protects fragile ecosys-tems — including forests, river systems, soils, and ocean fisheries — while, if possible, restoring them.

Nations whose subsistence farmers still form a significant proportion of the over-all population, though in recent decades termed "underdevel-oped," may in fact have some advantages in the post-growth world. Rather than continuing with ruinous attempts to install fuel-guzzling food and transport systems, these countries should be adopting the "appropriate" or "intermediate" technologies that have been advocated for several de-cades by E. F. Schumacher and others. Appropriate technology (or AT) is typically labor- and knowledge-intensive rather than capital-, resource-, and energy-intensive. Examples include the use of local natural materi-als for building, the small-scale generation of local power from methane digesters, and the purification of water in households with porous ceramic filters.

People in currently wealthy nations may well find themselves adopting similar technological strategies in the decades ahead. In the process, there may be a substantial reversal of the trend, seen since the beginning of the Industrial Revolution, toward greater wealth inequality among nations.

BOX 5.5 Human Scale Development

According to the school of "Human Scale Development" developed by Manfred Max-Neef, Antonio Elizalde, and Martin Hopenhayn, fundamen-tal human needs are ontological (stemming from the condition of being human); they are also few, finite, and classifiable — as distinguished from the conventional notion of economic "wants" that are infinite and insa-tiable.[46] They are also constant through all human cultures and through-out history; what changes is the set of strategies by which these needs are satisfied. Human needs are a system — that is, they are interrelated and interactive. The school of Human Scale Development is described as, "focused and based on the satisfaction of fundamental human needs, on the generation of growing levels of self-reliance, and on the construction of organic articulations of people with nature and technology, of global processes with local activity, of the personal with the social, of planning with autonomy, and of civil society with the state."[47]

Max-Neef classifies the fundamental human needs as:

- subsistence,
- protection,
- affection,
- understanding,
- participation,

- leisure,
- creation,
- identity, and
- freedom.

The Post-Growth Struggle Between Rich and Poor

If current levels of wealth inequality among the world's nations cannot be maintained in a non-growing, energy-starved economy, that doesn't necessarily mean that we are headed toward a future of perfect equity. Rather, the end of growth is likely to lead, in many instances, to a sharply heightened struggle between rich and poor for control of the world's vanishing wealth.

As GDP per capita has increased during the past two centuries, so has inequality in the distribution of wealth, both among nations and often within nations as well. These trends will not be abandoned without a fight.

The most widely used metric of economic inequality is the Gini coefficient, developed by Italian statistician Corrado Gini in 1912. A value of 0 reflects total equality, while a value of 1 shows maximal inequality. In the world currently, Gini coefficients for income within nations range from approximately 0.23 (Sweden, with the lowest level of economic inequality) to 0.70 (Namibia, with the highest). The US weighs in at 0.45—between Cote d'Ivoire at 0.446 and Uruguay at 0.452 (some agencies arbitrarily shift the decimal point two places to the right for all scores, giving the US a score of 45 instead of 0.45).

Inequality *among* nations can also be tracked with the Gini coefficient; it turns out that, in recent decades, the richest countries have pulled ahead while the poorest countries fell further behind, with a few in the middle (including China and India) playing a rapid game of catch-up. However, "catching up" has meant increasing wealth inequality within those "developing" nations. Current research shows that global income inequality peaked in the 1970s when there was little overlap between "rich" and "poor" countries. Since then, the rapid industrialization of nations like China, India, Indonesia, and Malaysia has complicated the picture. [48]

The absolute number of people living in poverty—across a range of definitions—has consistently declined globally during the past 50 years, and the percentage of people living in poverty has fallen even faster.[49] Nevertheless, according to a study by the World Institute for Development Economics Research at United Nations University, the richest one percent of adults has continued to pull ahead, owning 40 percent of global assets in the year 2000, with the wealthiest ten percent of adults accounting for 85 percent of the world total. The bottom half of the world adult population owns barely one percent of global wealth.[50]

The reasons for change in wealth inequality within and among nations are varied: tax policies, capital investment, culture, education, natural resources, trade, and history all play roles. Moreover, there is controversy between those who say inequality within nations is good because it stokes more growth (governments should aim for equality of opportunity, not equality in incomes, according to free-market advocates), and those who say too much income inequality is inherently unfair and tends to become structural and to foreclose economic opportunity for the majority of the world's people.

The end of growth will no doubt alter the prospects of both rich and poor, in both absolute and relative terms. Those with privilege will no doubt struggle to maintain it, while the poor, driven to desperation by generally worsening economic conditions, may in increasing numbers of instances organize or even revolt in order to increase their share of a shrinking pie.

In her 2008 book *The Shock Doctrine: The Rise of Disaster Capitalism*, Canadian anti-globalization author and activist Naomi Klein argued that modern neo-liberal capitalism thrives on disasters, in that politicians and corporate leaders take advantage of natural calamities and wars to ram though programs for privatization, free trade, and slashed social spending—programs that are inherently unpopular and would have little chance of adoption in ordinary times.[51] Klein's thesis seems confirmed in the present instance: the end of growth is presenting societies with an ongoing economic crisis, and we have already seen how, in the US, well-heeled investors and executives have benefited from government bailouts while millions of workers have lost jobs and homes. Austerity programs in Greece and Ireland have resulted in a similar upward distribution of rewards and downward spreading of sacrifices.

Prior to the financial crisis of 2008 some countries were already seeing a backlash by the poor against the kinds of economic predation that Klein highlights. Within Latin America, the Bolivarian Alliance for the Americas was initiated in 2004 with bilateral agreements between Venezuela and Cuba; the Alliance now numbers eight nations — including Ecuador, Nicaragua, and Bolivia — which are in the process of introducing a new regional currency, the *sucre*, to be used in place of the US dollar. The sucre is intended to serve as the common virtual currency of the Alliance for now, and eventually to become a hard currency. The Alliance aims for social welfare, bartering, and mutual economic aid rather than trade liberalization led by Washington. Bolivia's nationalization of its hydrocarbon assets and Ecuador's declaration of the illegitimacy of its national debt (because it was contracted by prior corrupt and despotic regimes) can be interpreted as expressions of these nations' rejection of the "shock doctrine" of the global economic elites.

Meanwhile, as we saw in Chapter 2, several European countries are seeing increasingly vocal popular opposition to the austerity plans being imposed by the EU and the International Monetary Fund in response to sovereign debt crises. For example, in November 2010, Ireland received an EU bailout accompanied by a requirement for cuts in domestic social spending. The latter provoked street protests in Dublin by tens of thousands of citizens. A general strike ensued, and there was talk of a change of government and failure of the budget.

For the US, recent history suggests that for the time being popular resistance to efforts by wealthy to shift bailout costs onto the middle and poor classes is likely to be muted. In the 1970s, corporations and wealthy individuals began a process of political organization by pooling funds and investing them in think tanks and media outlets. The number of registered Washington lobbyists rose from 175 in 1971 to nearly 2,500 in 1982.

Money flooded into political campaigns as never before, mostly favoring business-friendly, anti-tax candidates. The trend has only accelerated in recent years (today Washington lobbyists number almost 35,000).

At least partly as a result, the income gap between the richest and poorest Americans has grown. In 2010, according to the Gini index, US income inequality reached its highest level since the Census Bureau began tracking household income in 1967. The US has also achieved the highest disparity in household incomes among Western industrialized nations.[52]

This situation is largely accepted as a given by most Americans, and seldom discussed in the media except by marginalized leftist intellectuals. The super rich are more often idolized as role models than scrutinized for their methods of wealth accumulation. The payout of enormous Wall Street bonuses at a time when hundreds of billions in taxpayer dollars were being used to bail out the street's biggest firms drew expressions of dismay, but few organized protests. In short, there appears to be little basis within the nation for the mounting of an effective citizen-led movement to reduce economic inequality in the face of worsening financial conditions. In December 2010, Congress extended Bush-era tax cuts for the wealthiest Americans, signaling political leaders' disinterest in reducing inequality by way of the tax structure — the quickest and most effective means to that end. Altogether it would appear that, in the US, the economic consequences of the end of growth (unemployment, homelessness, and hunger) are likely to be viewed by most members of the poor and middle classes as evidence of personal failure; whatever anger comes from suddenly losing a job or from being unable to find one is likely to be directed against irrelevant scapegoats provided by politicians.

At the end of the previous section it was suggested that poor nations with large self-sufficient rural populations may fare relatively well in adapting to the end of growth. Such will not be the case with poor nations that depend largely on foreign aid. As food surpluses and loan packages disappear, many citizens in these nations are likely to be left without access even to the essentials of life. Victims of natural disasters in already aid-dependent nations will face especially dire conditions, and (as argued in Chapter 3) the number and intensity of natural disasters will almost certainly increase in the years and decades ahead. The current situations in Haiti, which is still struggling to recover from the earthquake of January 2010, and Pakistan, where millions continue to suffer from the effects of that year's floods, give some indication of just how abject life could be for many tens of millions of humans in the near future if efforts are not invested now in helping poor nations become more self-sufficient and resilient.

In the first three chapters we examined evidence that world economic growth is ending. In Chapter 4, we saw why market mechanisms that many assume can keep growth on track in the near future, if not indefinitely, are in fact incapable of doing so. This chapter paints a fairly dismal picture of a post-growth world characterized by heightened geopolitical and demographic competition, in which hundreds of millions who are currently enjoying or aspiring to a middle-class lifestyle may sink into poverty, and in which many millions more who are already poor may lose access to the barest elements of survival.

It is natural for readers to find this distressing. I may seem to have gone out of my way to focus relentlessly on negative prospects, discounting possibilities and opportunities while highlighting limits and dire outcomes.

It has been necessary to frame the issues this way because the end of growth is an inherently unattractive notion, and so most people are likely to avoid considering it, deny evidence that it is occurring, and fail to contemplate its implications, unless presented with an airtight case in its favor. The end-of-growth argument therefore has to be made carefully, thoroughly, even somewhat redundantly. But it must be made. If the observation that growth is ending is in fact valid, and if policy makers and citizens don't see or understand that economic expansion is no longer possible, they will continue to assume the impossible — that growth can and will continue indefinitely. In doing so they will increasingly be operating in a delusional state. People who are deluded this way may do things that make no sense in terms of the actual economic environment that is emerging, and will likely fail to do things that could help themselves and others adapt to new conditions. Opportunities will be wasted and human suffering will be increased unnecessarily.

Readers may think that by asserting that humanity can't continue to expand global economic activity the author is in effect saying that people aren't smart or innovative. That would be an incorrect interpretation. Humans certainly are adaptable and ingenious. However, there are always constraints on what we can do. Engineers take practical limits (such as the tensile strength, compression limits, and melting points of various materials and the load limits of structures) into account in every project they undertake. A smart, creative engineer can maximize performance within relevant limits, creating masterpieces in efficiency.

Inventors can reveal entirely new worlds in which former limits are transcended, but then new sets of limits come into play. For example, the Wright brothers overcame previous constraints on human mobility — and put succeeding generations of aviation engineers to work maximizing lift-to-drag ratios.

No matter how much creativity we bring to the table, blind disregard of limits can lead to disaster.

Humankind has been seduced by a temporary abundance of cheap fossil energy into ignoring limits to Earth's resources, and limits to the ability of economies to keep piling up debt. Our task now is to understand these limits so we can intelligently and inventively back away from them and begin to maximize our opportunities within them.

In the next two chapters we will explore how humanity might apply creativity to the task of adapting to depleting resources and stagnant or shrinking economies. As we are about to see, it is essential that we deal with the immediately looming monetary-financial crisis if we are to buy time to set ourselves on a course for a happier, more sustainable, and more secure future.

CHAPTER 6

MANAGING CONTRACTION, REDEFINING PROGRESS

*Only a crisis — actual or perceived — produces real change.
When the crisis occurs, the actions that are taken depend
upon the ideas that are lying around. That, I believe, is our
basic function: to develop alternatives to existing policies, to
keep them alive and available until the politically impossible
becomes politically inevitable.*

— Milton Friedman (economist)

Many analysts who focus on the problems of population growth, resource depletion, and climate change foresee gradually tightening constraints on world economic activity. In most cases the prognosis they offer is for worsening environmental problems, more expensive energy and materials, and slowing economic growth.

However, their analyses often fail to factor in the impacts to and from a financial system built on the expectation of further growth — a system that could come unhinged in a non-linear, catastrophic fashion as growth ends. Financial and monetary systems can crash suddenly and completely. This almost happened in September 2008 as the result of a combination of a decline in the housing market, reliance on overly complex and in many cases fraudulent financial instruments, and skyrocketing energy prices. Another sovereign debt crisis in Europe could bring the world to a similar precipice. Indeed, there is a line-up of actors waiting to take center stage in the years ahead, each capable of bringing the curtain down on the global banking system or one of the world's major currencies. Each derives its destructive potency from its ability to strangle growth, thus setting off chain reactions of default, bankruptcy, and currency failure.

The likely outcomes of a non-linearity response of the monetary-financial system to the end of growth thus constitute a wall in our path. Beyond the wall are other challenges and opportunities — challenges like oil depletion and climate change, and opportunities to reshape the economy so as to make it more sustainable over the long run, and to make it better serve human needs.

The depletion of resources and the buildup of greenhouse gases are gradual processes, though their various impacts will be subject to tipping points and will provoke short-term crises. Efforts to deal with these problems — such as building low-energy transport infrastructure and low-carbon food systems — will take a generation or more. That kind of time just won't be available to us if we can't get past the financial-monetary wall. If we hit the wall at full speed, our options will be severely and suddenly reduced. The economy, and society as a whole, may undergo an abrupt, dramatic, and chaotic simplification as trade virtually ceases.

So far, we are on course for full-force collision. The fundamental problems with our monetary and financial systems have not been addressed, but only papered over.

Our financial-monetary system is not just vulnerable to periodic internal disruptions like credit crises, it is inherently unsustainable in the emerging context of energy and resource constraints. And if the financial-monetary system seizes up, this will imperil society's ability to respond to any and all other crises. This means that, whatever our other priorities may be, we must also immediately devote effort to reforming the financial-monetary system.

This chapter is mostly about what governments can do — must do, in fact — to get past the wall of looming financial-monetary collapse. As we will see, there may be more than one strategy that could work. But having averted immediate collision, we won't be in the clear: this short-term barrier in humanity's path must be negotiated in a way that also steers us around slower-developing problems such as climate change and resource depletion. If not, civilization will carom from one crisis to the next.

The Default Scenario

Making economic forecasts i. always hazardous, as was pointed out in Box I.3 in the Introduction, "The Perils of Predication." Nevertheless it may be useful to outline a potential default scenario — one way that events could unfold if we continue on our current track. Things need not play out this way, but if we do nothing to alter our current trajectory they very well could.

If consumer spending fails to recover, so that demand for new loans continues to remain low, this will put pressure on major banks' balance sheets, making the toxic assets still on their books more difficult to conceal among what would otherwise be sounder, newer loans and investments. Unemployment will almost certainly remain high in the US (according to nearly all official forecasts), causing tax revenues to remain low and forcing drastic cuts in the budgets of cities, counties, and states. Other sovereign nations with high debt levels will be remain vulnerable to currency and credit crises. Central banks and some national governments (principally the US and Germany) will be compelled to extend more bailouts.

In effect, the global economy will be stuck with trillions of dollars in IOUs that cannot be repaid. And the number will continue to increase if policy makers continue to demand economic growth, because governments' attempts to restart growth will require the further expansion of claims in the form of debt.

Under these circumstances, national governments and central banks (including the IMF, which acts somewhat as a global central bank) will be the only entities capable of keeping banking systems, and hence the global economy as a whole, functioning. Governments and central banks will be acting under the assumption that they are merely priming the pump of the economy until conventional consumer-driven growth resumes. But as growth fails to revive, one intervention after another will be required — propping up major banks, guaranteeing hundreds of billions of dollars' worth of mortgages, or bailing out "too-big-to-fail" businesses. The result will be an incremental government takeover of large swaths of national economies, with central banks assuming more of the functions of commercial banking, and national governments underwriting production and even consumption.

In the US, this process will be enormously complicated by politics. One of the two main political parties is making resistance to expansion of government spending the centerpiece of its platform. Yet, whether Democrats or Republicans hold power in the US, the solution hit upon will eventually be more or less the same (recall: it was Republican President George W. Bush who extended the first round of bailouts and stimulus packages) — though the path toward achieving it is likely to be extremely contentious and littered with casualties. With states, counties, and municipalities nearing bankruptcy, the Federal government's hand may be forced: It must eventually either bail them out or permit the unfolding of a fiscal and human crisis that could spread to engulf the nation.[1]

The US government's expanding role in the economy is likely to be accompanied by greater reliance on the military for attempted solutions to national and international problems, for the following reasons. Cutting military spending will be problematic in a flagging economy, as that would create even more unemployment; substantial spending cuts in this area would likely be politically contentious in any case. Meanwhile, burgeoning trade, currency, and resource conflicts are likely to provoke saber-rattling responses from US adversaries and allies alike. Policy makers in a few strategic countries will be quick to back up tough talk with further investments in arms (though in many other nations military investment rates will fall for lack of available funds). The military may also be seen as the ultimate guarantor of domestic order.

Economic and demographic strains cannot help but stoke widespread dissent and unrest.[2] In response, governments are likely to become more repressive. In the US, again whether Republicans or Democrats are in power, this could mean increased surveillance, controls over the Internet, tightening laws governing freedom of expression, and sharp reductions in guarantees of civil rights and liberties — most likely in the name of protection from terrorism and in response to worsening natural disasters. Wikileaks aside, secrecy will be rampant — with the biggest secret of all being that leaders have no viable long-term strategy to stop the economy's slide.[3]

With support services (in the U.S: Social Security, Medicare, public schools, the food stamp program) stretched beyond their limits, we could see more public resentment against immigrants, especially in border states. Of course, the economic pain gripping the United States will not actually be the fault of immigrants — or China, Muslims, environmentalists, or even terrorists. Nor is the essential problem Big Government: As we have seen, the desperate effort to inflate government spending and power is more of an effect than a cause of the nation's predicament. The search for scapegoats will accomplish nothing, but it will consume enormous amounts of effort and produce needless casualties. A sound case can be made that bankers and government officials played key roles in the financial crisis, and these individuals should be held to account. But correctly assigning blame will not make the crisis go away.

Events could continue to play out along these lines for several years, with gradually worsening outcomes. Nationalization of the economy will not constitute a solution to society's difficulties; it will merely be a reflexive means of averting immediate meltdown. The phrase "bailout fatigue" has already entered the lexicon of policy makers, and will be the subject of increasing worry and controversy in coming years. Even with ballooning deficits and enormous spending programs, economic problems will only fester. Budget-balancing austerity measures will succeed only in reducing economic activity further. In either case, as energy becomes ever less affordable, economic productivity will decline and costs of long-distance trade will rise.

At some point in the next few years, stock and real estate values will plunge, banks will close, and businesses will shutter their doors. Monetary, financial, and social systems built upon the expectation of growth will simply fail in growth's absence. In the worst instance, that failure could take the form of a nearly complete cessation of trade, as occurred nationally in Argentina in December, 2001. Some sort of new economy would inevitably emerge from the wreckage, but in scale and scope it would be a shadow of the one we knew just a few years ago. Measured in GDP, it might correspond to the world economy of fifty, a hundred, or even a hundred and fifty years ago.

The pursuit of the ideals of fairness, openness, and freedom, and the fights against corruption, greed, and tyranny will of course continue, as they must, but these struggles will play out within the constraints of a shrinking economy. Promises of plenty if only new leaders and policies are put in place will prove hollow. Social progress could yield relative change in economic conditions (advancing the prospects of the poor versus the rich), but not absolute change (the economy will still be contracting); meanwhile, the more intense the conflict, the more resources will be consumed that might have been devoted to helping households and communities adapt.

Sadly, the scenario I have just laid out is not necessarily the worst-case outcome. It is possible to imagine ones in which environmental disasters or energy shortages play more prominent roles, and where collapse comes sooner and is more complete.

Whether contraction is chaotic or controlled, and whether it comes sooner or later, a radical simplification of the economy is more or less inevitable, as systems designed for cheap energy and economic growth slam up against environmental limits. And the risk of uncontrolled, chaotic collapse is considerable. As the 2010 Bundswehr (German military) report on Peak Oil put it: "A shrinking economy over an indeterminate period presents a highly unstable situation which inevitably leads to system collapse.... The risks to security posed by such a development cannot even be estimated."[4]

I am about to argue, however, that *economic contraction need not entail catastrophe and sorrow if the process is managed well.*

Haircuts for All...or Free Money?

To get past the wall of potential financial-monetary collapse, governments would have to resort to extraordinary emergency measures. In the best instance, this would create time and space to begin coming up with long-term, infrastructural responses to declining energy supplies and climate change — responses involving the redesign of transport systems, power generation and transmission systems, food systems, and so on. Of course, there is no guarantee that time, once gained, will be well spent. Nevertheless, in principle the wall can be traversed.

The essence of the wall is this: We have accumulated too many financial-monetary claims on real assets — consisting of energy, food, labor, manufactured products, built infrastructure, and natural resources. Those claims, essentially IOUs, exist in the forms of debt and derivatives.

Our debt cannot be fully repaid: every dollar saved in the past is owed ever-multiplying returns in the future, yet the planet's stores of resources are finite and shrinking. Claims just keep growing while resources keep depleting — and real prices of energy and commodities have begun rising. At some point it will become clear that this vast ocean of outstanding claims will never be honored, and the result could be a tidal wave of defaults and bankruptcies that would sweep away most of the economy.

In theory, as Harvard economic historian Niall Ferguson points out, there are six ways of resolving a debt crisis: (1) increasing the rate of GDP growth; (2) reducing interest rates; (3) offering bailouts; (4) accepting fiscal pain — reductions in benefits and standard of living; (5) injecting more money into the economy; or (6) accepting defaults, "including every type of non-compliance with the original terms of the debt contract."[5] If the premise of this book is correct and it has become nearly impossible to grow GDP, then we can eliminate option (1). Interest rates (2) cannot realistically be reduced lower than zero, which is essentially where they are now (for banks — though credit card interest rates are still in the range of 20 percent). As the debt problem worsens, bailouts (3) become more expensive and less effective. The austerity option (4) is distasteful to everyone and can only be pursued aggressively at the risk of a breakdown of social cohesion. Government printing of money (5) is frowned upon by trading partners and inflates away savings. Option (6), widespread defaults, could lead to a broad-scale failure of the monetary-financial system, so it will likely be avoided, except in limited circumstances.

Currently governments are dithering with all of these options, applying them in an *ad hoc* and piecemeal fashion. However, two of the six have at least a theoretical capability of being implemented in a fairly dramatic, strategic way if and when the crisis becomes otherwise unmanageable. These strategies would consist of a modified debt jubilee (a form of default, option 6), or a bout of inflation through the creation of non-debt-based currency (option 5). Both would come with major risks, but either could, in principle, buy time for the implementation of a more fundamental reform of the entire economic system.

A modified debt jubilee could take the form of a universal "haircut" — a term currently being used in financial circles to describe a situation where the market value of securities being held by financial firms as part of their net worth is significantly reduced. In the strategy being proposed, the "haircut" would apply to all financial claims. Government by edict would reduce all debt by a certain percentage — let's say, somewhere between 75 and 90 percent. At the same time, all investments and savings accounts above a certain figure (allowance would have to be made for pensioners and low-income individuals) would get the same treatment. The process would be complicated and unpopular, especially among those with the most to lose, but it might help get us past the wall. It would reduce economic activity significantly — that's going to happen anyway, even in the best instance — but it would also remove the overhang of debt that threatens to bring down the entire economy.

How might this work? Let's say, as a starting point, that we wanted to protect all assets below a certain level. In the US, perhaps all assets below $25,000 could remain untouched. Then, one simple way to administer the "haircut" would be to slice a decimal place off everyone's debts, savings, and other accounts. If you had a $250,000 mortgage, it would be knocked down to $25,000 — but your $20,000 savings account would survive unscathed, as it fell below the $25,000 limit designed to protect pensioners and other low-income individuals. Your debt overhang would have shrunk from $230,000 to $5,000. A wealthy person who had gained $5 billion through investing in hedge funds would now have only $500 million. A business that owed $750,000 in loans would now owe $75,000. And so on.

The net result would be a "re-set" in the relationship between claims and real assets, bringing that relationship back into a somewhat more workable balance. Of course, this "re-set" would be hugely controversial, confounding...and painful.

Sound far-fetched? Certainly, an action like this would not be undertaken unless other tactics had failed. It would yield winners and losers: although everyone would feel the effects, the impact would be uneven. At first glance, it seems those with the fewest assets and highest debts would suffer least. A more likely outcome would be widely distributed dislocations, unemployment, and so on, so there would be plenty of suffering to go around. But, this "re-set" would give us the opportunity — if we took advantage of it — to restructure our economic and financial systems to be more sustainable and resilient.

The second strategy would consist of governments or central banks creating debt-free money. This is how economist Richard Douthwaite, founder of the organization FEASTA and editor of the book *Fleeing Vesuvius*, describes it:

> The solution is to have central banks create money out of nothing and to give it to their governments either to spend into use, or to pay off their debts, or give to their people to spend. In the eurozone, this would mean that the European Central Bank would give governments debt-free euros according to the size of their populations. The governments would decide what to do with these funds. If they were borrowing to make up a budget deficit — and all 16 of them were in deficit in mid-2010, the smallest deficit being Luxembourg's at 4.2 percent — they would use part of the ECB money to stop having to borrow. They would give the balance to their people on an equal-per-capita basis so that they could reduce their debts, or not incur new ones, because private indebtedness needs to be reduced too. If someone was not in debt, they would get their money anyway as compensation for the loss they were likely to suffer in the real value of their money-denominated savings. Without this, the scheme would be very unpopular. The ECB could issue new money in this way each quarter until the overall, public and private, debt in the eurozone had been brought sufficiently down for employment to be restored to a satisfactory level.[6]

An alternative would be to pass laws against usury (for example, any interest rate greater than 20 percent would become illegal), then print enough money to accelerate inflation beyond 20 percent. People's debts would decline over time, as would the value of money being held. The government could spend money into existence for social welfare programs, thus ensuring that retirees and other vulnerable groups don't get hit too hard.

In the US, a version of the "free money" strategy is being advocated by Ellen Brown, author of *Web of Debt*. Brown argues that the United States Congress has the constitutional authority to coin money, but historically has needlessly delegated the power of money creation to the banking system—and, since 1913, to the Federal Reserve. The Federal government has on occasion created money directly, without borrowing—notably to finance the Civil War. The Federal Reserve's second bout of quantitative easing, in 2010, was essentially a version of this strategy: the Fed bought government debt with money created on the spot, and interest from the debt will be rebated to the Treasury. However, Brown argues that the best way to pursue this option would be for the government itself to directly issue debt-free money, rather than for the Fed to do it through a more circular means.

The objection usually raised against government "printing" of large amounts of new money is that this would be highly inflationary: the US economy could suffer the same fate as Weimar Germany, with its currency becoming virtually worthless and all savings being wiped out in the process. Brown disagrees:

> Adding money ("demand") to an economy with high unemployment and unused productive capacity serves to increase productivity, increasing goods and services or "supply." When supply and demand increase together, prices remain stable. And adding money to the money supply is obviously not hazardous when the money supply is shrinking, as it is now.... Financial commentator Charles Hugh Smith estimates that the economy now faces $15 trillion in writedowns in collateral and credit. If those estimates are correct, the Fed could, in theory, print $15 trillion and buy up the entire federal debt without creating price inflation. That isn't likely to happen, but it does make for an interesting hypothetical.[7]

Over the short run, emergency measures could include the Fed buying up short-term municipal bonds in order to ease state and county fiscal crises, and the European Central Bank doing something similar with bonds of member nations, creating money to fill the gap left by the contraction in the money supply which resulted from the financial crisis of 2008 and that has led to soaring budget deficits.[8]

Another related, longer-term measure that could help, according to Brown, is the establishment of state or provincial banks. Currently, North

Dakota has the only state-owned bank in the US, established by the state legislature in 1919. The bank's original purpose was to free farmers and small businesses from indebtedness to out-of-state bankers and railroad companies. By law, the state deposits all its funds in the bank, and deposits are guaranteed by the state. The Bank of North Dakota is a bankers' bank, partnering with private banks to loan money to farmers, real estate developers, schools, and small businesses. It also purchases municipal bonds.

What would be the advantage to the state of having of such a bank? With fractional reserve lending, banks extend credit (create money as loans) in amounts equal to many times their deposit base. If a state owns its bank, it need not worry about shareholders or profits, so it could lend to itself or to its municipal governments at zero percent interest.[9] If these loans were rolled over indefinitely, this would be essentially the same as creating debt-free money.[10]

Clearly, none of these strategies can solve the long-term problems of declining energy and minerals, rising population, and worsening environmental crises. They are merely ways to avert the looming wall of monetary-financial collapse. Once we have bought some time, we must begin to redesign certain basic structures of the economy that currently function properly only in a context of constant growth.

One of these structures consists of the money we use.

Post-Growth Money

Over the past few centuries, with urbanization and the expansion of trade, money has become essential to the functioning of societies. Today even the most remote human communities depend on the vagaries of a technology of exchange that only a few people seem to understand, and that periodically suffers crises of over- or under-valuation.

As we saw in Chapter 1, metal-based forms of money were dominant for more than a millennium until the 20th century, when virtually the entire world adopted debt-based currencies no longer backed by metal. Yet, due to the requirement for interest payments, debt-based currency can only function well in an expanding economy.[11] As we've also noted, a transition back to metal-backed currencies is problematic. This means we will *have to* reinvent money in the years ahead.

Given money's importance, it would be natural to assume that the discussion about currency systems and how they function is robust, featuring a wide-ranging literature, numerous college courses devoted to the subject, and so on. This is far from being the case. The notion that alternative kinds of money may be possible, some of them superior to debt-based currency, has occurred to some (Henry Ford, for example, is reputed to have been enamored with the idea of an energy-backed currency), but only a very few thinkers seem to have explored money systems in depth.[12] The project requires that we start by taking account of genuine human needs and then ask how money can be used to help satisfy them. It also requires a careful examination of our current monetary system and its vulnerabilities and failures. The goal should be to design a system — possibly involving several interlocking currencies to be used simultaneously but for different functions — that would liberate the exchange process from political and banking interests that presently tend to distort and commandeer it for their own ends: a system that could serve the needs of a steady-state or shrinking economy as easily as those of one that is growing.

In the previous section we noted the recommendation by Ellen Brown and Richard Douthwaite that governments (including US state governments and Canadian provincial governments) create their own debt-free money. Economist Herman Daly, author of *The Steady State Economy*, has mooted essentially the same idea: If there were a 100 percent reserve requirement for banks, this would get them out of the business of creating money; meanwhile government could award lump sums to new entrants into the economy (every 18-year-old would receive enough to pay for a college education) and lend money into existence at zero percent interest for socially worthwhile projects.[13]

Brown and Douthwaite insist that we don't need banks in order to create money, and money doesn't have to be loaned into existence with interest accruing. The kind of money we've been using for the past century is based on credit — which is helpful because it allows the money supply to expand or contract as conditions require.[14] After all, money is more of a relationship than a thing: it is, in Brown's words, "a legal agreement, a credit/debit arrangement, an acknowledgment of a debt owed and a promise to repay."

However, credit-money need not be dependent on bank-generated, interest-bearing debt, if buyers' and sellers' credit accounts can be cleared directly. The idea is simple and is already being demonstrated in hundreds of local currencies that are in active use around the world. Many are versions of Local Exchange Trading Systems (LETS), in which each transaction is recorded as a corresponding credit and debit in the two participants' accounts. The quantity of currency issued is always and automatically sufficient and does not depend on a bank or government for issuance.[15]

Some local currencies have no physical representation and consist only of the mutual credit of participants, while others use paper notes representing anything from a number of hours worked (Ithaca, NY, has a local currency known as "Ithaca Hours") to quantities of food in storage (Willits, CA, has "Grange Grains"). The backing of local currencies with something tangible seems useful primarily as a way of defining the unit of account, and as a way of making the currency more "real" and acceptable to users.

Herman Daly is one of several authors who have advocated for the proliferation of local currencies; others include Dierdre Kent (*Healthy Money, Healthy Planet*); Richard Douthwaite (*The Ecology of Money*); Bernard Lietaer (*The Future of Money: Creating New Wealth, Work and a Wiser World*); and Thomas Greco, Jr. (*Money: Understanding and Creating Alternatives to Legal Tender*).[16]

Among the most successful of US local currencies is BerkShares, traded in the Berkshire region of Massachusetts, launched in 2006.[17] BerkShares are available at five participating banks, where 95 Federal Reserve dollars may be exchanged for 100 BerkShares and then used to purchase goods and services on a one-to-one basis at over 400 businesses. Over 2.8 million BerkShares have been traded at banks since launch.

Complementary currencies are especially useful in situations where the national currency is, for whatever reason, failing to serve the needs of producers and consumers — as occurred during the Argentine economic collapse of 2001–2002, when small-denomination, interest-free provincial bond IOUs issued by local governments enabled trade to continue.

In his book *The End of Money and the Future of Civilization*, Thomas Greco, Jr. describes mutual credit clearing as both a fundamental monetary advance, and a solution to the basic and irreconcilable problems of debt-based money.[18] Through mutual credit clearing, participants in the economy are, in effect, creating their own currency as needed.

Bernard Lietaer cites as examples credit-clearing systems such as the Commercial Credit Circuits ("C3") in Brazil that enable small businesses to bypass banks for short-term financing; he points out that Uruguay allows payment of taxes in C3 currency.[19] Greco goes on to offer a regional development plan based on credit clearing as well as suggestions for a complete web-based credit-clearing trade platform.[20]

The unit of account used in credit-clearing exchange systems can vary. Greco notes possibilities including a basket of commodities, a unit of energy (such as the kilowatt hour), an existing currency unit (the US dollar), or a labor standard (a statistical unit of labor productivity). Even gold or silver could be used as a unit of account—though this would not require stockpiling of metals or actual payment of accounts in coin. There would be a considerable advantage to using the same unit of account globally so as to facilitate trade, but currencies themselves would work best as nested, diverse systems—with local, regional, and national currencies in simultaneous use.

Greco, Douthwaite, Daly, and others agree that governments should support and facilitate the emergence of local currencies. Lietaer notes that complementary currencies "meet unmet needs with unused resources," and uses the mileage programs of the major airlines as an example of complementary currencies already familiar to most people.

Historically, governments have used their monopoly on the issuance of currency as a way to consolidate state power; legal tender laws, which require citizens to use the national currency, are the primary means of maintaining this monopoly. Greco, echoing Austrian-School economist Friedrich von Hayek, advises rescinding legal tender laws, writing that, "There should be a strict separation between money and the state. Any financial instruments issued by the government must be made to stand upon their own merits in the financial markets."[21] If this principle were generally accepted, says Greco, inflation would cease to exist.[22]

Given a future of reduced global trade (because of scarcer fuel), and a greater need for resilience (which means more diversity, interconnectivity, and redundancy in basic societal support systems), local currencies would seem to make a great deal of sense. In practice, most local currencies in use during the past few decades have existed on the fringes of national economies, but this is largely because of legal tender laws maintaining the monopoly status of national currencies. As the global economy fails to grow, and as debt-based national currencies therefore become more dysfunctional, local currencies could enable commerce to continue.

For the past several decades the US dollar, created by commercial banks through interest-bearing loans, regulated by the Federal Reserve, and mandated by legal tender laws, has acted as a de facto global currency. It will take a financial-monetary earthquake to dislodge the dollar from that role and function. But pressure is building along the fault lines of the global economy, and even within the US political system. In recent months, US House of Representatives members Ron Paul (by some accounts the member furthest to the political right) and Dennis Kucinich (by some accounts furthest to the left) have both called for the abolition of the Fed; Paul advocates a return to the gold standard, while Kucinich backs the direct creation of debt-free money by the Federal government.

Meanwhile, the use of barter within the US is trending sharply upward.[23] Mutual credit-clearing exchanges and local currencies represent a significant advance over barter, but without the drawbacks of our present national debt-based currencies.

BOX 6.1   A Global Currency?

There is some tentative and controversial indication that policy makers at the highest levels are aware of the vulnerability of existing currencies and have begun thinking about alternatives. Some anti-globalization bloggers believe there may be an Orwellian solution in store: Point 19 of the official communiqué from the 2009 G20 summit noted, "We have agreed to support a general SDR allocation which will inject $250bn (£170bn) into the world economy and increase global liquidity." SDRs, or Special Drawing Rights, are "a synthetic paper currency issued by the International Monetary Fund." Could the IMF be testing the waters for the creation of a global currency?[24] Full implementation of a global currency would require many more steps, including the setting up of a full-fledged global central bank (the IMF is not currently equipped to fulfill this role, though perhaps it could be revamped for the purpose). Globalization critics fear that an IMF currency would not only inherit the reserve status of the US dollar, but could become be the primary means of exchange within all countries. Presumably, as a condition for inclusion into this new global currency system, nations would have to accept the kinds of austerity packages recently meted out to Greece and Ireland.

The evidence suggests that any plans that may be in the works for a global currency are far from implementation. Meanwhile, one can't help but wonder whether a better outcome could be achieved if the project of designing a new money system were undertaken with more transparency, and if alternatives such as direct credit clearing systems were taken into account.

Post-Growth Economics

The past three decades, and especially the past three years, have seen an explosion of discussion about alternative ways of thinking about economics. There are now at least a score of think tanks, institutes, and publications advocating fundamentally revising economic theory in view of ecological limits. Many alt-economics theorists question either the possibility or advisability of endless growth.

The fraternity of conventional economists appears to be highly resistant to these sorts of challenging new ideas. Governments everywhere accept unquestioningly the existing growth-based economic paradigm, and this confers on mainstream economists a sense of power and success that makes them highly averse to self-examination and change. Therefore the likelihood of alternative economic ideas being adopted anytime soon on a grand scale would seem vanishingly small. Nevertheless, alternative thinking is still useful, because as growth ends the managers of the economy will sooner or later be forced to try other approaches, and it will be extremely important to have conceptual tools lying around that, in a crisis, could be quickly grasped and put to use.

As noted in Chapter 1, conventional economics starts with certain basic premises that are clearly, unequivocally incorrect: that the environment is a subset of the economy; that resources are infinitely substitutable; and that growth in population and consumption can continue forever. In conventional economics, natural resources like fossil fuels are treated as expendable income, when in fact they should be treated as capital, since they are subject to depletion. As many alternative economists have pointed out, if economics is to stop steering society into the ditch it has to start by reexamining these assumptions.[25]

The following four fundamental principles must be established at the core of economic theory if economics is to have any relevance in the future:

- Growth in population and consumption rates cannot be sustained.
- Renewable resources must be consumed at rates below those of natural replenishment.
- Non-renewable resources must be consumed at declining rates (with rates of decline at least equaling rates of depletion), and recycled wherever possible.[26]
- Wastes must be minimized, rendered non-toxic to humans and the environment, and made into "food" for natural systems or human production processes.[27]

Further, economics must aim for a dynamic balance between efficiency (maximizing throughput) and resilience (adaptability, redundancy, diversity, and interconnectivity)—whereas today economists focus almost entirely on efficiency.[28]

The contributions of the alternative economists (via schools of thought known as ecological economics, environmental economics, and biophysical economics) can be divided into three broad categories: critiques of existing economic system, proposals for an alternative system, and strategies for making the transition from one to the other.[29]

In his book *Prosperity Without Growth*, British economist Tim Jackson writes: "During the [period since 1950] the global economy has grown more than 5 times," and economists expect it to quadruple again by mid-century. "This extraordinary ramping up of global economic activity has no historical precedent," according to Jackson.

> It's totally at odds with our scientific knowledge of the finite re-
> source base and the fragile ecology on which we depend for sur-
> vival.... Questioning growth is deemed to be the act of lunatics,
> idealists and revolutionaries. But question it we must. The idea of
> a non-growing economy may be an anathema to an economist.
> But the idea of a continually growing economy is an anathema to
> an ecologist.... The only possible response to this challenge is to
> suggest — as economists do — that growth in dollars is "decoupled"
> from growth in physical throughputs and environmental impacts.
> But...this hasn't so far achieved what's needed. There are no pros-
> pects for it doing so in the immediate future. And the sheer scale
> of decoupling required...staggers the imagination.[30]

The New Economics Foundation in London recently published a book-length study titled *Growth Isn't Possible*, which asks whether goals related to mitigating climate change can be met in the context of continued global economic growth. Its conclusion: "Economic growth in the OECD cannot be reconciled with a 2, 3, or even 4°C characterization of dangerous climate change."[31]

Herman Daly, one of the pioneers of ecological economics (he published *Toward a Steady State Economy* in 1973 and *Beyond Growth* in 1996, and co-authored a textbook titled *Ecological Economics* in 2004), differentiates between *economic* growth and *uneconomic* growth.[32] For Daly, uneconomic growth *consists of GDP gains that are accompanied by static or declining social benefits, as for example when a certain amount of short-term growth is achieved by undermining ecosystems whose services have a greater long-term value.*[33]

In Europe, a "degrowth" movement has taken root, founded on the ideas of Mohandas Gandhi, Leopold Kohr, Jean Baudrillard, André Gorz, Edward Goldsmith, Ivan Illich, and Serge Latouche.[34] The work of Romanian economist Nicholas Georgescu-Roegen (1906–1994) was especially pivotal in setting the movement on its path: his 1971 book titled *The Entropy Law and the Economic Process* pointed out that neoclassical economics fails to acknowledge the second law of thermodynamics by not accounting for the degradation of energy and matter. Georgescu-Roegen's thinking had in turn been influenced by that of chemist-turned-economist Frederick Soddy (1877–1956), author of *Wealth, Virtual Wealth and Debt* (1926), which sought to bring economics into line with the laws of thermodynamics and which critiqued fractional-reserve banking.[35] The French translation of Georgescu-Roegen's book in 1979 under the title *Demain la décroissance* ("Tomorrow, Degrowth") spurred *décroissance* thinking and organizing that eventuated in the first International Degrowth Conference in Paris in 2008 and the founding of a French-language newspaper, *La Décroissance: Le journal de la joie de vivre*, published in Lyons.

In the United States, the term "degrowth" is seldom mentioned; however, over the past twenty years a similar trend in thinking has spurred the "voluntary simplicity" movement, which questions the environmental, psychological, and social costs of ever-growing consumption. The movement has roots in the ethical beliefs of religious groups like the Amish, but also in the writings of philosopher Henry David Thoreau (1817–1862) and back-to-the-land pioneers Scott and Helen Nearing (1883–1983; 1904–1995, authors of *Living the Good Life*).[36] The books *Voluntary Simplicity* by Duane Elgin (1981), and *Your Money or Your Life* by Joe Dominguez and Vicki Robin (1992), and the documentary film "Affluenza" (1997) helped define this movement, which now also features magazines and newsletters to assist in the formation of local simple living networks.[37] Many simplicity advocates promote Buy Nothing Day, which falls on the Friday following Thanksgiving Day in the United States, as an antidote to pre-Christmas shopping frenzy.

The US has also spawned systematic critiques of standard economic theory. Henry George (1839–1897) has been called America's most important home-grown economist; his writings explored the implications of the principle that each person should own what he or she creates, but that everything found in nature, most importantly land, should belong equally to all humanity.[38] Economist Thorstein Veblen (1857–1929) criticized the wastefulness of consumption for status.[39] More recently, the book *Small*

Is Beautiful by German-British economist E. F. Schumacher (1911–1977) inspired Bob Swann (an American pioneer of land trusts) to found the E. F. Schumacher Society, which is now the New Economics Institute, one of several US organizations that promote a basic restructuring of the economy according to ecological principles.[40]

If growth is impossible to sustain, what alternative goal should economies pursue? Herman Daly (who was a student of Georgescu-Roegen) has for nearly three decades advocated a "steady-state economy," which he describes as "an economy with constant stocks of people and artifacts, maintained at some desired, sufficient levels by low rates of maintenance 'throughput,' that is, by the lowest feasible flows of matter and energy from the first stage of production to the last stage of consumption."[41] A steady-state economy would aim for stable or mildly fluctuating levels in population and consumption of energy and materials; birth rates would equal death rates, and saving/investment would equal depreciation.

The goal of a steady-state economy is now being actively promoted by the Center for the Advancement of a Steady State Economy (CASSE), headquartered in Arlington, VA, with chapters elsewhere in the country.[42] The president of the organization, Brian Czech, is author of *Shoveling Fuel for a Runaway Train* (2000).[43]

In his 2007 book *Managing Without Growth*, Canadian economist Peter Victor presents a model of the Canadian economy that shows "it is possible to develop scenarios over a 30 year time horizon for Canada in which full employment prevails, poverty is essentially eliminated, people enjoy more leisure, greenhouse gas emissions are drastically reduced, and the level of government indebtedness declines, all in the context of low and ultimately no economic growth."[44]

Some critics of the steady-state economy concept have assumed that keeping consumption constant would require harsh government controls. However, Daly and others contend that such an economy could flourish in the context of a constitutional democracy with a common-sense mixture of markets and market regulations. Markets would still allocate resources efficiently, but some vital decisions (such as permissible rates of resource extraction and the just distribution of resources, especially those created by nature or by society as a whole) would be kept outside the market.

A few nations and communities are already moving in the direction of a steady-state economy. Sweden, Denmark, Japan, and Germany have arguably reached situation in which they do not depend on high rates of growth to provide for their people. This is not to say these countries have only smooth sailing ahead (Japan in particular is facing a painful adjustment, given its very high levels of government debt), but they are likely to fare better than other nations that have high domestic levels of economic inequality and that have gotten used to high growth rates.

Sweden is now home to a number of eco-municipalities. Inspired by economist Torbjörn Lahti and by Karl-Henrik Robèrt, founder of the Natural Step Movement, these formerly depressed industrial towns have made an official and deliberate commitment to "dematerialize" their economies and to foster social equity.[45] Övertorneå, Sweden's first eco-municipality, saw a 20 percent unemployment rate during the recession of the early 1980s and lost 25 percent of its population (prior to becoming an eco-municipality), but now boasts a thriving ecotourism economy based on organic farming, sheepherding, fish farming, and the performing arts. The town has reached its 2010 goal of being a free of fossil fuels. Hällefors, a former steel town that also suffered from high unemployment 20 years ago, now has an economy based on renewable energy, organic farming, and culinary arts. Other eco-communities exist in Norway, Finland, and Denmark.[46]

For the world as a whole, the transition from a growth-based economy to a steady-state economy is likely to be far more problematic than the examples in the preceding paragraph might suggest: ecotourism will never be the economic backbone of New York, Beijing, or Mumbai—though organic farming will likely be the main engine for a growing number of smaller communities.

Which raises the question: How do we get there from here? Aside from creating non-debt-based currencies (as discussed above), what strategies could help ease the way toward a healthy post-growth world economy? Herman Daly and other steady-staters advise policies along the following lines:

- A cap–auction–trade (or cap-and-dividend) system for extraction rights for basic natural resources;

- A shift away from taxing income and toward taxing resource depletion and environmental pollutants;
- Limits on income inequality;
- More flexible workdays; and
- The adoption of a system of tariffs that would allow countries that implement sustainable policies to remain competitive in the global marketplace with countries that don't.

One of the fundamental problems with markets, acknowledged by nearly all economists, is the tendency for businesses to externalize costs ("externalities," in economic theory, are costs or benefits from a transaction that are not reflected in the price). For example, companies that burn fossil fuels—thereby releasing air pollutants—typically pass the resulting health bills and clean-up costs on to nearby communities, or the nation, or the world as a whole. It is possible to *internalize* such costs through laws and regulations. One strategy is to collect "Pigovian" taxes from businesses equal in amount to their negative, externalized costs to society. Another solution is to define property rights more carefully (e.g., the right of residents in a community to clean air and water) so that efforts to remedy violations of those rights carry legal weight. Many conventional economists believe that such measures will solve the problem of externalities without need for government intervention in markets, but Herman Daly and Josh Farley have argued that in reality such measures are only partly effective, as the interests of future generations are still not taken into account.[47] One remedy that Daly and Farley suggest is making the rights of future generations to certain resources, such as to the ecosystems responsible for generating life-support functions, explicit and inalienable.[48]

Henry George championed the idea of a "single tax" on the use of land (while accepting private ownership of land, he advocated the public capture of all value it generates), with the proceeds shared by society; this was a purist solution to the problems of economic inequality, monopolies, and environmental externalities. A partway measure in this direction consists of levying high taxes on land values. Pittsburgh, PA, did this in 1913 by instituting a high tax on unimproved land held for speculation, and as a result land values there have remained far more stable than in other cities.[49] If the government captures any increases in land values, it eliminates speculative demand for land, thus avoiding speculative bubbles and keeping land cheaper for non-speculative uses. Land equity partnerships and land trusts (including agricultural land trusts) are other proven ways to overcome the landlord-tenant dilemma and remove land from the speculative market.[50]

Futurist Hazel Henderson, author of *Ethical Markets: Growing the Green Economy*, advises governments to charge a financial transaction tax of one percent or less.[51]

> This would not affect the trades of 99.9 percent of all Americans. But it would put a major crimp in the games that the big boys play. Let the quants use their brainpower to cure cancer rather than to craft complex computerized trading systems that leave society with less than nothing. A small transaction tax could generate over a $100 billion a year from Wall Street — and in the process, bring those ridiculous bonuses and profits back in line with the real economy.[52]

Henderson also advocates breaking up too-big-to-fail banks and businesses and fostering non-profit community development finance institutions (CDFIs) to address the capital needs of micro-businesses.

To discourage trans-border financial capital flows that exploit the labor and resources of less-industrialized countries, Daly calls for downgrading the IMF and World Bank into mere clearinghouses that collect fees from countries that run both surpluses and deficits in their current and capital accounts. Daly would also remove price barriers to "non-scarce" intellectual capital — including royalty payments to patent holders. Such barriers often prevent less-industrialized countries from developing the renewable energy technologies necessary to bypass fossil fuels.

One final requirement in the transition from a growth economy to a steady-state economy is the reform of corporate law. Corporations enable individuals to pool financial resources to pursue commercial interests under a legal structure that limits liability for employees and investors. In the US, corporations also enjoy the status and rights of legal persons. In effect, this gives them the financial resources to influence public policy, and to exploit people and nature, without moral or legal responsibility. In fact, corporate officers are virtually required by law to place value to shareholders above all other considerations. University of British Columbia law professor Joel Bakan describes the modern corporation as "an institutional psychopath"; in the documentary "The Corporation" (2003), he claims that if the behavior of corporations were ascribed to ordinary people, the latter would be considered to exhibit the traits of antisocial personality disorder. In that same film, former Republican Party candidate for Senate from Maine, Robert Monks is seen remarking: "The corporation is an externalizing machine, in the same way that a shark is a killing machine."[53] The environmental ethic inherent in the corporate legal structure could be summarized as: "Use resources as fast as possible until they're gone."

Alternative economists argue that the genuine benefit of corporations (their ability to pool capital to achieve socially useful purposes) could be better achieved through cooperatives — which have a long history of success. Credit unions are cooperative banks; some utilities operate as cooperatives; and there are also housing, manufacturing, and agricultural cooperatives.[54] The following seven principles are central to the cooperative movement:

1. Voluntary and open membership,
2. Democratic member control,
3. Member economic participation,
4. Autonomy and independence,
5. Education, training, and information,
6. Cooperation among cooperatives, and
7. Concern for community.[55]

Cooperatives have the potential to avert overuse of resources by placing other values, including the interests of future generations, ahead of profit. Indeed, the organization "Coop America," which began as a sort of cooperative of US cooperatives, in 2009 changed its name to "Green America."

Gross National Happiness

After World War II, the industrial nations of the world set out to rebuild their economies and needed a yardstick by which to measure their progress. The index soon settled upon was the Gross National Product, or GNP—defined as the market value of all goods and services produced in one year by the labor and property supplied by the residents of a given country. A similar measure, Gross Domestic Product, or GDP (which defines production based on its geographic location rather than its ownership) is more often used today; when considered globally, GDP and GNP are equivalent terms.

GDP made the practical work of economists much simpler: If the number went up, then all was well, whereas a decline meant that something had gone wrong.

Within a couple of decades, however, questions began to be raised about GDP: perhaps it was *too* simple. Four of the main objections:

- Increasing self-reliance means decreasing GDP. If you eat at home more, you are failing to do your part to grow the GDP; if you grow your own food, you're doing so at the expense of GDP. Any advertising campaign that aims to curb consumption hurts GDP: for example, vigorous anti-smoking campaigns result in fewer people buying cigarettes, which decreases GDP.
- GDP does not distinguish between waste, luxury, and a satisfaction of fundamental needs.
- GDP does not guarantee the meaningfulness of what is being made, bought, and sold. Therefore GDP does not correlate well with quality of life measures.
- GDP is "*Gross* Domestic Product"; there is no accounting for the distribution of costs and benefits. If 95 percent of people live in abject poverty while 5 percent live in extreme opulence, GDP does not reveal the fact.[56]

In 1972, economists William Nordhaus and James Tobin published a paper with the intriguing title, *Is Growth Obsolete?*, in which they introduced the Measure of Economic Welfare (MEW) as the first alternative index of economic progress.[57]

Herman Daly, John Cobb, and Clifford Cobb refined MEW in their Index of Sustainable Economic Welfare (ISEW), introduced in 1989, which is roughly defined by the following formula:

ISEW = personal consumption + public non-defensive expenditures – private defensive expenditures + capital formation + services from domestic labor – costs of environmental degradation – depreciation of natural capital

In 1995, the San Francisco-based nonprofit think tank Redefining Progress took MEW and ISEW even further with its Genuine Progress Indicator (GPI).[58] This index adjusts not only for environmental damage and depreciation, but also income distribution, housework, volunteering, crime, changes in leisure time, and the life-span of consumer durables and public infrastructures.[59] GPI managed to gain somewhat more traction than either MEW or ISEW, and came to be used by the scientific community and many governmental organizations globally. For example, the state of Maryland is now using GPI for planning and assessment.[60]

During the past few years, criticism of GDP has grown among mainstream economists and government leaders. In 2008, French president Nicholas Sarkozy convened "The Commission on the Measurement of Economic Performance and Social Progress" (CMEPSP), chaired by acclaimed American economist Joseph Stiglitz. The commission's explicit purpose was "to identify the limits of GDP as an indicator of economic performance and social progress." The commission report noted:

What we measure affects what we do; and if our measurements are flawed, decisions may be distorted. Choices between promoting GDP and protecting the environment may be false choices, once environmental degradation is appropriately included in our measurement of economic performance. So too, we often draw inferences about what are good policies by looking at what policies have promoted economic growth; but if our metrics of performance are flawed, so too may be the inferences that we draw.[61]

In response to the Stiglitz Commission there have been increasing calls for a Green National Product that would indicate if economic activities benefit or harm the economy and human well-being, addressing both the sustainability and health of the planet and its inhabitants.[62]

One factor that is increasingly being cited as an important economic indicator is *happiness*. After all, what good is increased production and consumption if the result isn't increased human satisfaction? Until fairly recently, the subject of happiness was mostly avoided by economists for lack of good ways to measure it; however, in recent years, "happiness economists" have found ways to combine subjective surveys with objective data (on lifespan, income, and education) to yield data with consistent patterns, making a national happiness index a practical reality.

In *The Politics of Happiness*, former Harvard University president Derek Bok traces the history of the relationship between economic growth and happiness in America.[63] During the past 35 years, per capita income has grown almost 60 percent, the average new home has become 50 percent larger, the number of cars has ballooned by 120 million, and the proportion of families owning personal computers has gone from zero to 80 percent. But the percentage of Americans describing themselves as either "very happy" or "pretty happy" has remained virtually constant, having peaked in the 1950s. The economic treadmill is continually speeding up due to growth and we have to push ourselves ever harder to keep up, yet we're no happier as a result.

Ironically, perhaps, this realization dawned first not in America, but in the tiny Himalayan kingdom of Bhutan. In 1972, shortly after ascending to the throne at the age of 16, Bhutan's King Jigme Singye Wangchuck used the phrase "Gross National Happiness" to signal his commitment to building an economy that would serve his country's Buddhist-influenced culture. Though this was a somewhat offhand remark, it was taken seriously and continues to reverberate. Soon the Centre for Bhutan Studies, under the leadership of Karma Ura, set out to develop a survey instrument to measure the Bhutanese people's general sense of well-being.

Ura collaborated with Canadian health epidemiologist Michael Pennock to develop Gross National Happiness (GNH) measures across nine domains:

- Time use
- Living standards
- Good governance
- Psychological well-being
- Community vitality

- Culture
- Health
- Education
- Ecology

Bhutan's efforts to boost GNH have led to the banning of plastic bags and re-introduction of meditation into schools, as well as a "go-slow" approach toward the standard development path of big loans and costly infrastructure projects.

The country's path-breaking effort to make growth humanly meaningful has drawn considerable attention elsewhere: Harvard Medical School has released a series of happiness studies, while British Prime Minister David Cameron has announced the UK's intention to begin tracking well-being along with GDP.[64] Sustainable Seattle is launching a Happiness Initiative and intends to conduct a city-wide well-being survey.[65] And Thailand, following the military coup of 2006, instituted a happiness index and now releases monthly GNH data.[66]

Michael Pennock now uses what he calls a "de-Bhutanized" version of GNH in his work in Victoria, British Columbia. Meanwhile, Ura and Pennock have collaborated further to develop policy assessment tools to forecast the potential implications of projects or programs for national happiness.[67]

Britain's New Economics Foundation publishes a "Happy Planet Index," which "shows that it is possible for a nation to have high well-being with a low ecological footprint."[68] And a new documentary film called "The Economics of Happiness" argues that GNH is best served by localizing economics, politics, and culture.[69]

No doubt, whatever index is generally settled upon to replace GDP, it will be more complicated. But simplicity isn't always an advantage, and the additional effort required to track factors like collective psychological well-being, quality of governance, and environmental integrity would be well spent even if it succeeded only in shining a spotlight of public awareness and concern in these areas. But at this moment in history, as GDP growth becomes an unachievable goal, it is especially important that societies re-examine their aims and measures. If we aim for what is no longer possible, we will achieve only delusion and frustration. But if we aim for genuinely worthwhile goals that *can* be attained, then even if we have less energy at our command and fewer material goods available, we might nevertheless still increase our satisfaction in life.

Policy makers take note: Governments that choose to measure happiness and that aim to increase it in ways that don't involve increased consumption can still show success, while those that stick to GDP growth as their primary measure of national well-being will be forced to find increasingly inventive ways to explain their failure to very *un*happy voters.

Our Problems Are Resolvable In Principle

We've just seen how the economy could be put on the right track. But sorting out the economy is not enough to save the world; that would be just the first step.

The world's environmental dilemmas are likewise amenable to resolution, at least in principle. As support for that statement one can point to piles of "how-to-save-the-world" environmental articles and books—in fact I can point to *literal* piles of such books here in my little home office. Which suggests a way to approach writing this section of the book: rather than painstakingly assembling a balanced overview of an immense and wide-ranging literature, perhaps all that's really needed is for me to look around and grab a few titles off the shelves.

The first one that comes to hand is Lester Brown's *Plan B 4.0: Mobilizing to Save Civilization*.[70] In some ways we need go no further: Brown has provided a masterful overview of the world's 21st-century threats (oil and food security, rising temperatures and rising seas, water shortages, etc.) and the ways to contain or overcome them—by eradicating poverty, conserving resources, reforming the world's food system, raising energy efficiency, and developing renewable energy. There it is, folks: that's all you need to know. Just go out and do it. (Brown's very latest book, *World on the Edge: How to Prevent Environmental and Economic Collapse*, which I didn't have at the time of this writing, appears to be an updated and improved version of *Plan B*.)[71]

Ah, but how could we stop with just one book? Next in the stack is one I couldn't resist: my own largely neglected previous volume, *The Oil Depletion Protocol: A Plan to Avert Oil Wars, Terrorism and Economic Collapse*. It outlines a simple framework for guiding world policy regarding oil — and, in principle, all other non-renewable natural resources. Since we know that we cannot continue increasing rates of extraction forever, it makes sense to conserve such resources by deliberately reducing extraction rates now. If we did this in a coordinated way, we could keep resource prices from fluctuating destructively, reduce the incentive for nations to compete for dwindling supplies, and help jumpstart the inevitable transition to renewable alternatives.[72] What's not to like about that?

A third book that comes easily to hand is Albert Bates's *The Biochar Solution: Carbon Farming and Climate Change*. Bates has long been a prophet regarding climate change and is a veteran organic farmer; in this book he provides an excellent overview of a widely-researched technique for removing carbon from the atmosphere while building soil — a win-win solution if ever there was one.[73]

But wait — there are some problems we haven't addressed. How about transportation in an oil-constrained future? Take a look at *Transport Revolutions: Moving People and Freight Without Oil*, by Richard Gilbert and Anthony Perl, or *An American Citizen's Guide to an Oil-Free Economy* by Alan Drake, a veteran proponent of rails as being far more efficient than highways.[74] The problem of overpopulation must be mentioned again here — but we have already discussed the admirable and effective work of Population Media Center in Chapter 5; for more on solutions, see Bill Ryerson's chapter "Population: The Multiplier of Everything Else" in *The Post Carbon Reader*.[75] Conflict resolution methods and new governance models are covered in Roy Morrison's *Ecological Democracy*.[76] And the crisis in biodiversity is addressed an article in a recent issue of *Solutions* magazine — "Facing Extinction: Nine Steps to Save Biodiversity," by Joe Roman, Paul Ehrlich, et al.[77] The looming crisis in the world's food systems is tackled in a report I co-wrote with Mike Bomford a couple of years ago, "The Food and Farming Transition."[78]

I could keep going. The list of critical problems facing civilization is nearly endless, but each one of those problems has been addressed with proposals and model projects aimed at mitigating it. These are the tools we want to have lying around as crisis hits, though they'll only be useful if we actually pick them up and learn to wield them.

This chapter began with a rather dark "Default Scenario," yet we went on to see that there are solutions to the economic problem it portrayed; moreover, as we've just noted, there are potential answers to all our other critical problems as well. If civilization fails, it won't be for a lack of good ideas. Some of these have been around since the 1970s — a few since the 1870s. Which brings up the question: Why, if so many solutions are available, does my "default scenario" for the future look so dreary?

Perhaps the suggestion that "Our problems are resolvable in principle" needs to be followed by an "if" clause and a "but" clause.

The "if" clause: *If we are willing to change our way of life and the fundamental structures of society.* Many people assume that solving our problems means being able to continue doing what we are doing now. Yet it is what we are doing now that is creating our problems. Every "solution" mentioned above comes at a cost in terms of fundamental changes in individual and societal behaviors and priorities.

The "but" clause: *But our society as a whole is not inclined to do what is required to solve them, even if the consequences of failing to do so are utterly apocalyptic.* This statement seems bizarre on its face. Who would prefer to see economic collapse, the exhaustion of precious natural resources, the disappearance of millions of species, the failure of food systems — and resulting misery and death for millions upon millions of humans? Well, no one, if we put it that way. Yet the choices are not always so clear-cut, and we humans are hard- and soft-wired with genetic and psychological programming that can make it very difficult for us to undertake costly short-term behavioral change in order to avert future catastrophe.[79]

It may be cynical to say that policy makers will do the right thing only after all other alternatives have been exhausted.[80] But for the solutions we have been discussing, this does seem to be more or less the case. And this is true not just of policy makers, but the majority of us worker bees as well.

Paul Ehrlich and Robert Ornstein made a pioneering effort to understand our species' inability to pre-respond to impending, foreseeable crises in their book *New World New Mind* (1989), which describes the mismatch between the human nervous system and the complexities of our modern world.[81] While early hunter-gatherers evolved quick reflexes to cope with immediate threats in a limited environment, people in modern industrial societies face long-range problems not readily apparent to the five senses — growing population, climate change, resource depletion, and proliferation of debt. At their cores, our fight-or-flight brains just aren't up to dealing with these kinds of slowly developing dilemmas, even though our more advanced cerebral faculties enable us to define both challenge and potential solutions.

A more recent book, *American Mania: When More Is Not Enough*, by neuroscientist Peter Whybrow, digs deeper and reflects more recent research.[82] Whybrow notes that evolution equipped us to seek status and novelty, and to engage in conspicuous consumption. In our species' past, there were perfectly good reasons for these tendencies: they helped us survive and achieve reproductive success. But today, in a world of over-consumption, they keep us locked into behaviors that actually undermine our survival prospects.

Within our brains, dopamine plays a key role in governing motivation and stimulating the senses of reward and pleasure. On the primordial savanna, we got a hit of dopamine every time we discovered a tasty root or bagged a prey animal; today, stock trading lights up the same brain circuitry. But what helped us survive in one situation imperils us in the other. On the savanna, our early ancestors always needed the next meal, and then the next, and so the dopamine response evolved to be transitory. But today this means that, for the stock trader, no amount of profit is ever "enough." When we modern urbanites get a dopamine "hit" from a new car, a bigger house, or an end-of-year bonus, we may know intellectually that Earth simply can't keep supplying us with ever-increasing flows of such goodies, but it's hard to stop. We may even say we are "addicted" to shopping or some other aspect of consumption — but what we are really addicted to is the feeling it gives us.

According to Whybrow, Americans are particularly susceptible be-cause they are descended from immigrants with a higher frequency of the "exploratory and novelty-seeking D4-7 allele" in the dopamine receptor system; these immigrants, after all, were individuals who were willing to cross an ocean to pursue opportunity. Americans, he argues, are therefore disproportionately prone to impulsivity and addiction. Whybrow doesn't condemn Americans, whom he describes as "a self-selected group of hard-working opportunists with an insatiable hunger for self-improvement"; he merely points out that consumerism got its start in the US for reasons that have to do with biology as well as history.

Addiction is closely related to habituation: repeated use of an addictive drug typically leads to higher levels of tolerance. The same is true of dopamine-generating activities. Withdrawal from those activities leads to lower dopamine levels, so continuous acclimation to those activities is required to keep dopamine at normal levels, while a higher "dose" of activity is needed to get achieve the "high" that came the first time around. In his article "The Psychological and Evolutionary Roots of Resource Over-consumption Revisited," Energy blogger and former hedge-fund manager Nate Hagens writes:

> After each upward spike, dopamine levels recede, eventually to below the baseline. The following spike doesn't go quite as high as the one before it. Over time, the rush becomes smaller, and the crash that follows becomes steeper. The brain has been fooled into "thinking" that achieving that high is equivalent to survival (even more so than with food or sex, which actually *do* contribute to survival) and the "consume" light remains on all the time. Eventually, the brain is forced to turn on a self-defense mechanism, reducing the production of dopamine altogether — thus weakening the pleasure circuits' intended function. At this point, an "addicted" person is compelled to use the substance not to get high, but just to feel normal — since one's own body is producing little or no endogenous dopamine response. Such a person has reached a state of anhedonia, or inability to feel pleasure via normal experiences.[83]

Just as our brain circuitry can addict us to overconsumption, it also keeps us from responding to slowly accumulating environmental threats. Hagens points out that our brains are adept at calculating risks and rewards, and at applying discount rates according to the timing of events. We give the present predominantly more weight than the future when making decisions: an immediate reward is worth more to us than one promised next year, and an immediate threat will provoke more avoidance effort than one certain to emerge down the line. So even though the cost of averting climate change (in terms of loss to GDP) would be less than the eventual cost of climate change itself, we are generally unwilling to pay that smaller, immediate cost.

The limits to our ability to change behavior to avert crisis come not just from our individual brain wiring but from the psychology of organizations. While people within organizations individually have the characteristics we have been discussing, organizations themselves tend to develop their own defenses again change.

Political organizations, for example, tend to foster a culture in which insiders (politicians) are encouraged to tell outsiders (the people) what the latter want to hear, while withholding information about problems that cannot be solved without substantial sacrifice, or problems that cannot be blamed on other, competing politicians.

Both political and commercial organizations tend to elevate short-term priorities. For corporations, quarterly profits are the prime motivator, while politicians make decisions based on the next election cycle. Ironically, however, absent an immediate military threat, government policy tends to evolve very slowly, regardless of the urgency of the environmental or economic issues facing it.[84]

On top of all this, there are entrenched interests — people and institutions that profit from the system the way it is, don't want to give up those profits, and have the means to shape policy and public opinion. This is hardly a trivial point: billions of dollars are spent strategically in lobbying and public relations by corporations and wealthy individuals with the goal of shaping, delaying, or eliminating environmental legislation or reforms of the financial industry.

All of this would seem to suggest that human beings are simply incapable of conserving resources and that we are genetically wired to use the planet up and drive ourselves to extinction. But that's not entirely true. There are countervailing human tendencies exemplified in the traditions of indigenous peoples who made decisions based on the likely impacts on the seventh generation yet to come. Traditional societies planned ahead, made a virtue of thrift, and in many cases even held voluntary poverty as an ideal.[85]

These kinds of cultural values evolved slowly in response to environmental limits. During the past two centuries of rapid economic growth, such values have tended to be lost and forgotten. Peter Whybrow explains why:

> Selfish behaviors are reward-driven and innate, wired deeply into the survival mechanisms of the primitive brain, and when consistently reinforced, they will run away to greed, with its associated craving for money, food, or power. On the other hand, the self-restraint and the empathy for others that are so important in fostering physical and mental health are learned behaviors — largely functions of the new human cortex and thus culturally dependent. These social behaviors are fragile and learned by imitations much as we learn language.[86]

All of the solutions to our growth-based problems involve some form of self-restraint. That's why most of those solutions remain just good ideas. That's also why we will probably hit the wall, and why the outcomes described in the previous chapters of this book are likely. The sustainability revolution *will* occur. The depletion of nonrenewable resources ensures that humankind will eventually base its economy on renewable resources harvested at rates of natural replenishment. But that revolution will be driven by crisis.

The crucial question is, how serious will that crisis have to be to get our collective attention and force us to change our behavior? Will the crisis be so severe as to destroy the very basis of civilization? If so, we will have lost everything worthwhile that human beings have achieved during our past few centuries of struggle, invention, and inquiry. It need not be so, and by working now to ensure that the tools that are needed to enable the economy and society to adapt to the post-growth era are sharpened and available, we can create the conditions for a rapid response when our collective internal discounting mechanisms finally adjust to the scale of the crisis facing us.

Nevertheless, if some scale of impact is inevitable, this poses profound immediate challenges for individuals, families, and communities. How should we be preparing?

7 LIFE AFTER GROWTH

There is more to life than increasing its speed.
— Mohandas K. Gandhi

I made the decision to write this book with some trepidation. After all, the world economic system is held together largely by the belief and faith that it will continue to grow. It's a confidence scheme, in the purest sense. Publishing a book arguing that growth is now effectively impossible (except in limited instances) undermines the belief, faith, and confidence that bind investors, lenders, and borrowers together in a functioning system. It makes a financial-currency crash more likely, and the victims of such a crash would include not just bankers, but nearly everyone.

My staying quiet could help buy time for us all. But there is no assurance that the time so purchased would be used to fix the problems we have been discussing. In fact, governments and central banks have already bought time — several trillion dollars' worth — but are doing very little to make fundamental changes to our economic system so that it can function in the post-growth era.

Our collective global conversation about the economy needs to change. We need to be thinking and talking about how to adapt to the end of growth. I don't know how to help catalyze that conversation without first pointing out some inconvenient facts — starting with the fact that our economy currently is set up to fail under the kinds of circumstances that are unfolding around us (resource depletion and catastrophic environmental decline). If political leaders and voices in the major media are unwilling to consider the possibility that growth is ending, then at least this information should be available to receptive individuals and communities so they can prepare themselves for what is coming. This was the argument *for* publication.

Even so, there is irony and risk. The strategies that individuals should be pursuing to prepare for the end of growth (disengaging from consumerism, getting out of debt, becoming more self-sufficient) are things that—if everyone did them—would keep the economy from recovering and would push us further into recession.

When the short-term interests of the economy conflict with the long-term interests of communities, a majority of individuals, and the natural world, we have a dilemma on our hands. In some respects, it is not an entirely new one (this conflict was implicit in Marx's critique of capitalism), but it is becoming acute and more difficult to hide. Resolving the conflict in favor of the economy is no solution when individuals, communities, and nature are imperiled to the point that economic growth cannot continue in any case—which is exactly the situation we face. Resolving the conflict entirely in favor of individuals is no solution if this results in a substantial reduction in the integrity of the social bonds the economy knits together: that is, if we are reduced to a random collection of seven billion humans, each scrambling for survival in the absence of functioning currencies and governments. In that case, the result would be universal chaos, confusion, and suffering.

Somehow we have to prepare individually for the ending of growth (a process likely to be accompanied by economic and political upheavals) while at the same time preserving and building social cohesion and laying the groundwork for a new economy that can function in a post-growth, post-fossil fuel environment. It's a tall order, but nothing less will do.

Setting Priorities

As someone who has for several years been speaking and writing about the consequences of impending energy scarcity, I'm often asked for personal advice. "Where should I live in order to avoid the worst impacts from Peak Oil?" "What career should I prepare myself for?" "What should

I invest in?" I'm generally uncomfortable answering such questions. I'm no prophet, merely a trend spotter. The trends I see are broad and deep, but the details of their unfolding could be surprising to everyone, myself certainly included.

Nevertheless, these are legitimate questions, and it is possible to extend some general advice. After long consideration, I've decided to put that advice for individual adaptation to the end of growth on a website (TheEndofGrowth.com) rather than include it in this book. That way it can be frequently updated as I hear from readers.

At the same time, it's important to recognize that there is only so much that individuals can do on their own. Some of the disruptions we may be facing would not be of short duration. A few weeks' worth of stored food and water, though essential, will be of only temporary help. Over longer time frames, our most valuable personal assets will be functioning local communities composed of people who, despite their differences, are willing and able to work together to solve problems and maximize opportunities.

The maintenance of social cohesion must be our single highest priority in a future of mounting economic and environmental challenges.

The challenge of building or maintaining community solidarity will be greater in some places than others. Rebecca Solnit's book *A Paradise Built in Hell* cites examples showing that in crisis people often re-discover community and what is intrinsically important in life.[1] However, Lewis Aptekar's *Environmental Disasters in Global Perspective* adds layers of complexity: People's responses to crisis seem to depend on the duration of the crisis, whether it can be blamed on other people, and on pre-crisis social and economic conditions.[2]

During the past few decades North Americans created a way of life in which people moved frequently, saw their homes as investments rather than just as places to live, and learned to ferry children around by van and SUV to soccer games and ballet lessons rather than encouraging them to spontaneously organize their own outdoor pastimes. The result: throughout the vast, sprawling suburbs of the US and Canada, most people simply don't know their neighbors. Any of them. At all. This is a bizarre situation, and it will probably be a dangerous one in the case of crisis.[3]

It's hard to emphasize this point sufficiently: Get to know your neighbors. These may be people with whom you share very little in terms of politics, religion, or cultural interests; that fact is beside the point. When push comes to shove, these are people you may need to depend on. Find ways — perhaps innocuous ones at first, such as a discussion about pruning a common shade tree or the sharing of surplus summer garden veggies — to make contact and to begin to build trust.

The remainder of this chapter is devoted to suggestions for what you can do to help your community become more resilient and better able to weather the approaching storms.

Transition Towns

Given the looming energy and environmental threats outlined in this book, it's evident that something like the following is called for. We need a grassroots movement that educates people about these challenges and helps them develop strategies to reduce their dependence on fossil fuels. It should aim to build community resilience, taking account of local vulnerabilities and opportunities. Ideally, this movement should frame its vision of the future in positive, inviting terms. It should aim to build a cooperative spirit among people with differing backgrounds and interests. While this movement should be rooted in local communities, its effectiveness would increase if it were loosely coordinated through national hubs and a global information center. The work of local groups should include the sharing of practical skills such as food production and storage, home insulation, and the development and use of energy conserving technologies. The movement should be non-authoritarian but should hold efficient meetings, training participants in effective, inclusive decision-making methods.

That may sound like a tall order. But here's some good news: that movement already exists. It's called Transition Initiatives, and communities that have one of these initiatives often call themselves Transition Towns.[4] The "transition" that's being referred to is away from our current growth-based, fossil-fueled economy and toward a future economy that is not only sustainable but also fulfilling and interesting for all concerned.

Transition Initiatives got their start in 2005 in Britain through the work of a Permaculture teacher named Rob Hopkins. In his *Transition Handbook*, Hopkins tells how he came up with the strategy, and sets forth a range of useful guidelines for groups.[5] Nearly all of Rob's prose is saturated with irrepressible optimism:

> Transition Initiatives are not the only response to peak oil and climate change; any coherent national response will also need government and business responses at all levels. However, unless we can create this sense of anticipation, elation and a collective call to adventure on a wider scale, any government responses will be doomed to failure, or will need to battle proactively against the will of the people.... Rebuilding local agriculture and food production, localizing energy production, rethinking healthcare, rediscovering local building materials in the context of zero energy building, rethinking how we manage waste, all build resilience and offer the potential of an extraordinary renaissance — economic, cultural and spiritual.[6]

Hopkins is careful to call Transition a "research project"; in a "cheerful disclaimer" on the Transition website he points out that there is no guarantee of success, because what is being attempted is unprecedented.

> We truly don't know if this will work. Transition is a social experiment on a massive scale. What we are convinced of is this:
> - if we wait for the governments, it'll be too little, too late
> - if we act as individuals, it'll be too little
> - but if we act as communities, it might just be enough, just in time.[7]

Hopkins lives in the old market town of Totnes in the southwest of England; with a population of 7,444, it is the most advanced of all Transition Towns.[8] There, over 30 projects have started and nine themed groups meet regularly to discuss food, buildings and housing, arts, transport, and education, among other topics. In 2009, as a result of Transition efforts, Totnes was awarded a grant of £625,000 for a program called "Transition

Streets," a street-by-street approach to energy efficiency, community building, and domestic micro-generation. Totnes now has its own local currency, as well as a *Renewable Energy Society that is charged with owning and profitably running the renewable energy generating capacity of the region.* The Totnes Food Hub is a co-operative, member-owned alternative food distribution system; members can order fresh food from local producers at affordable prices and have it delivered, ready for collection, to a convenient location in the center of town. Transitioners also host clothes swaps, based on the idea that most of us have in our closets newish clothes that we never wear and that others may be able to use. One of Transition Totnes's biggest accomplishments was the development of a town-approved Energy Descent Action Plan—a multi-decade staged plan for reducing dependence on fossil fuels in all significant areas (transport, food, home heating, etc.).

After the successful "unleashing" of Transition Totnes in 2006, the idea spread rapidly (though the pace seems to have leveled off in the past year); there are now 350 recognized Transition Initiatives in over 30 countries, with about 80 in the US (and about 150 more groups in America now forming). A team of trainers travels the globe offering help in getting Initiatives started, and thousands of people in over a dozen countries have taken the two-day Transition Training.

In Whidbey, WA, the local Transition Initiative features a Local Economy Action Group (a community think tank for creating a sustainable economy on Whidbey Island), a Clean Energies Cooperative (that focuses on alternative-fueled transportation), a Whidbey Citizens Climate Lobby, a Local Food Action Group (with subgroups that map the island's food resources, glean and distribute surplus fruit, and prune trees for better yields), a bi-weekly discussion group on Alternative Building, and a support group for people who want to discuss how the economic crisis is impacting them.

There are limits and obstacles to the Transition strategy. In the worst instance, Transition can manifest as merely another talk shop for lefties and aging former hippies. However, Hopkins recognizes that it must be something very different from this if it is to succeed, and that Transition must address practical matters having to do with infrastructure and practical economics. In a recent essay he noted:

The infrastructure required for a more localized and resilient future, the energy systems, the mills, the food systems and the abattoirs, has been largely ripped out over the past 50 years as oil made it cheaper to work on an ever-increasingly large scale, and their reinstallation will not arise by accident. They will need to be economically viable, supported by their local communities, owned and operated by people with the appropriate skills, and linked together.[9]

Hopkins went on to list the various infrastructure elements required to enable a town-sized economy function. At least Transition sees what's needed, even if it's not yet entirely up to the task.

Common Security Clubs

In addition to Transition Initiatives, something more is called for. As we work together on getting beyond oil and other fossil fuels, we also need to find mutually supportive ways to deal with immediate impacts from the fracturing of the economy. Joblessness, home foreclosures, and business failures are leaving a wake of destruction in communities, neighborhoods, and families. How are we to cope? Must we shoulder these losses household by household, or does it make more sense to get together with friends and neighbors to find shared ways to come to terms with the economic impacts of the end of growth?

Once again there is good news. A program already exists called Common Security Clubs with exactly this mandate.[10] The program was started by a team of economic justice and ecological transition activists connected to the Institute for Policy Studies and On the Commons, who put together a pilot curriculum in January 2009. Over 55 clubs have followed the suggested program, while another 100 or so groups have been inspired and informed by it and have adopted other names. The Clubs have a three-pronged strategy:

Learning together: Using popular education tools, videos and shared readings, participants deepen their understanding of economic issues and explore questions like: Why is the economy in distress? What are the ecological factors contributing to the economic crisis? What is our vision for a healthy, sustainable economy? How can I reduce my economic vulnerability? How can I get out of debt?

Mutual aid: Through stories, examples, web-based resources, a workbook, and mutual support, participants reflect on what makes them secure. How can I help both myself and my neighbor if either of us faces foreclosure, unemployment, or economic insecurity? What can we do together to increase our economic security?

Social action: Common Security Clubs recognize that many of our challenges won't be overcome through personal or local efforts. State, national, and even global economic reforms are needed. What state and federal policies will increase our personal security? How can we become politically engaged so as to further those policies? Many clubs, animated by "break up with your bank" and "move your money" reform efforts, have relocated personal, congregational, and other funds out of Wall Street and into local banks and credit unions.

The Common Security Clubs website offers tools for facilitators who want to start a group, as well as stories from existing Clubs.[11] One story is from a "Resource Sharing Group" in rural Maine started by Connie Allen.[12] Connie writes: "I knew several people who were living with limited income either because of unemployment, under-employment, retirement or voluntary simplicity. And I thought, if we put this group together, we could all benefit from it. It would make life easier for all of us."

And she was right. But what she didn't expect was how much fun they would have. "We used to meet in the basement of the local library, about twelve of us, each week," explains Connie. "The librarian was always asking us what we were laughing at. Somehow we just always had a lot of fun when we met. And we helped each other in all kinds of ways."

"We would bulk shop together," Connie says. "And we'd tell each other about sales and ways we had saved money and time each week."

The group shared lawn mowers, books, and tools; helped one member set up her new office; organized a yard and craft sale for forty people; set up a website to share information and list items to sell; offered tutorials in a variety of subjects; brainstormed job possibilities; met for potlucks; and shared inexpensive recipe ideas and savings tips.

They even kept an "emergency jar" at the center of the table. People would often put 50 cents or a dollar into it at meetings, though it wasn't required. The money didn't get used very often, but much like the group itself, it "provided a sense of security just in knowing it was there."

Common Security Clubs could gain effectiveness if they were supported by a national PR campaign — but who would pay for it? Certainly not the Federal government, which continues to spin the fiction that our national goal must be to return to a life of carefree motoring through anonymous suburbs. It is only as that ideal fades, along with the government's ability to continue bailing out banks, that necessity might conceivably lead to support for what is essentially a no-cost partial solution to burgeoning household financial crises.

Putting the New Economy on the Map

As important and helpful as Transition Initiatives and Common Security Clubs are, they share an annoying shortcoming: they tend to be invisible to the majority of people even in the towns and cities where they happen to be flourishing. Transition Los Angeles has been holding meetings since 2008; by all accounts it is a successful Initiative that is now spinning off a series of smaller and more localized chapters in the dozens of towns that make up the greater Los Angeles conurbation. Still, one wonders what proportion of the overall populace in that region has any awareness whatever of its existence: if the figure exceeds one percent, that would be pleasantly surprising. Even fewer Angelinos are likely to know they have the option of forming or joining a Common Security Club.

This suggests one more wrench may be needed in our post-growth toolbox — a way to make elements of the new post-growth economy visible and accessible to the community at large. The following is a strategy that you personally may not have the means to fully realize. But you could work with others to pursue it, and it is an idea that could be taken up by existing community organizations (including Transition Initiatives, Common Security Clubs, or Community Action Agencies).[13]

Why not rent a storefront and give the new economy a presence on Main Street? Here's the rationale. As America adjusts to the New Normal of tight credit, chronically less-affordable energy, high unemployment rates, rising levels of homelessness, and steeply declining tax revenues, strategies will be needed to help swelling ranks of low-income people adjust and adapt. National policies designed to ease credit, lower mortgage rates, or provide basic financial assistance (including extended unemployment benefits) may help over the short term, but over the longer term many needs will be better met locally by largely volunteer-driven non-profit organizations, co-ops, and hybrid public-private agencies and programs.

Many of these kinds of organizations already exist, but (like Transition Initiatives and Common Security Clubs) they are largely invisible. What's required is a way for them to become persistently recognizable. What better way than to plant them together right in the middle of town?

Even the earliest towns had a center, which was usually occupied by an open plaza where people could gather informally, a market, a ceremonial building, and a civic building of some kind. Everyone knew where the center of town was, and it was there that the life of the community came to focus. In many modern industrial cities (particularly in the US), the downtown has withered. Shopping malls, government complexes, and mega-churches are distributed throughout the city and its suburbs, all connected by hundreds of miles of highways. Nevertheless, the center of town still has symbolic and historic meaning, and as cheap transport fuel becomes a thing of the past city centers may regain their former importance.

If the new economy is to have much chance of taking root, it has to be planted in an identifiable location. Imagine the transformative potential of a loosely coordinated national network of locally-based Community Economic Laboratories (CELs), each equipped to help citizens solve practical problems arising during the breakdown of the old growth-based, fossil-fueled economy and the evolution of its replacement.[14]

The mission of a CEL would be to increase personal and community resilience by bringing together in one place the essential elements of a new local, resilient economy.

While for most citizens goods and services have traditionally been delivered by way of market relationships based on jobs and commercial interactions between individuals and for-profit businesses, even in good times some individuals occasionally (others chronically) require special assistance, which is usually provided by non-profit service agencies or government programs. In especially hard times, large numbers of individuals and families lose jobs and incomes, and therefore access to the goods and services that the market economy formerly provided them. At the same time, tax-starved governments are hard pressed to step in to make services available to rapidly expanding rolls of unemployed. At such a time, it could be helpful to explore new and innovative ways of fostering self-sufficiency through the coordination of a variety of cooperative, non-profit, market-based, and government-led ventures that spring from, and are adapted to, unique local conditions.

The CEL would be a local multi-function hub consisting of a number of independent organizations and businesses dedicated to helping people impacted by hard times, and to providing the armature around which a new economy can be woven. It would offer a variety of services, as well as opportunities for self-improvement, learning, enterprise incubation, and community involvement. Some possible examples of participating organizations and businesses:

- A food co-op
- A community food center, including commercial food-processing, food-preserving, and food-storage facilities available at low cost (or on labor-barter basis) to small-scale local producers[15]
- A community garden with individual beds available for seasonal rental, as well as communal beds growing produce for soup kitchens
- A health center offering free or inexpensive wellness classes in nutrition, cooking, and fitness
- A free (and/or barter) health clinic
- Counseling and mental health services
- A tool library, or an open-source customizable set of industrial machines[16]
- A work center that connects people who have currently unused skills with needs in the community — work can be compensated monetarily or through barter
- A legal clinic
- A credit union offering low-interest or even no-interest loans (on the model of the JAK bank in Sweden)[17]

- A recycling/re-use center that turns waste into resources of various kinds—including compost and scrap—and into re-manufactured or re-usable products
- A co-op incubator
- A local-currency headquarters and clearinghouse
- A local-transport enterprise incubator, possibly including car-share, ride-share, and bicycle co-ops as well as a public transit hub
- A shelter clearinghouse connecting available housing with people who need a roof—including rentals and opportunities for legal organized squatting in foreclosed properties, as well as various forms of space sharing
- A community education center offering free or low-cost classes in skills useful for getting by in the new economy—including gardening, health maintenance, making do with less, energy conservation, weather-stripping, etc.

Many communities already host one or more of these services, businesses, and organizations, but typically they are scattered throughout town. This is a disadvantage: individuals and families who have recently become jobless or homeless may be disoriented and less mobile, and therefore unable to access a variety of geographically dispersed opportunity centers. Commercial space in the downtown areas of many cities is already abundantly available due to the recession; if a CEL were able to obtain use of an iconic vacant building formerly housing a bank or department store, such an edifice would lend architectural validity to the efforts of community members to come together in providing for their neighbors.

Like a shopping mall, the CEL would be most successful if "anchored" by two or three substantial enterprises—such as a food co-op, community service organization, credit union, or transport co-op. One possible "anchor tenant" (or, in ecological terms, "pioneer species") would be a Sustainable Commercial Urban Farm Incubator (SCUFI) program, designed to train aspiring commercial urban farmers, assist with startup financing, help secure land, and provide them with technical and business support.[18]

Uniform national "branding" of CELs would be much less important than each community's sense of ownership of its unique, successful co-lab-

---

| BOX 7.1 | Investing in Sustainability: A Letter from an Eco-Entreprenur |

As stocks of non-renewable resources deplete and flow rates decline, society will need to better steward stocks and flows of renewable resources. My company was built around rebuilding stocks of topsoil while producing a commercial agricultural crop.

To finance this endeavor we formed an investment fund. This structure allows individuals or institutions to place their financial capital with us, which we use to purchase farmland and convert it organic. We recognize that not all land investments are equal, and that investments in conventional farmland where soil stocks are not being rebuilt will yield diminishing returns—and quickly so in the absence of commercial fertilizer inputs.

Profits, while important, are only one of many metrics with which to evaluate a business. We are certified as a "B-Corp," which provides a standardized way of measuring how business practices are supporting environmental and social values. (We're proud to have received the highest B-Score yet of 183.)

Unfortunately, funds such as ours are legally restricted by the Securities and Exchange Commission to "accredited" investors (i.e., individuals with at least $1 million net worth or $200k in annual income). However, anybody with a retirement account may be able to talk to whoever manages their money and ask them to place more emphasis on investments that rebuild natural capital instead of depleting it.

There are many ways to make a difference outside of the investment world. Reducing household expenses, developing a more self-reliant home economy, and reaching out to others in your community to share skills, equipment, and time in an informal way or through local currency systems are all rewarding options. If the financial system is going to become less reliable, then we will need to make other support systems more robust.

—*Jason Bradford (Manager, farmlandlp.com)*

oratory. Nevertheless, a national network could help quickly disseminate best practices, success stories, challenges, and other relevant information.

The CEL idea is not entirely new, and there are already several existing projects that have at least some of the characteristics described above:

- Social Innovation Center in Toronto (socialinnovation.ca)
- Working Centre in Waterloo, Canada (theworkingcentre.org/wscd/wscd_main.html)
- Bucketworks in Milwaukee, WI (bucketworks.org/about-bucket works)
- Springboard Innovation Centre in Torfaen, UK (springboardinnova tion.org.uk)
- The Hive in Portland, OR (leftbankproject.com/hive)
- The Plant in Chicago (plantchicago.com)
- Citizen Space in San Francisco (citizenspace.us)
- ShareExchange Project in Santa Rosa, CA (shareexchange.coop)

The last of these is in my town and just opened; it already hosts a Made Local Marketplace, the Sonoma County Timebank (an alternative currency), the Green Bough Health Cooperative, and the Work With Lounge (an entrepreneurial co-working space and micro-enterprise business incubator).

What Might a Sustainable Society Look Like?

Are these strategies sufficient to smooth our way through the economic and environmental crises of the next few decades? Unfortunately, no: It's going to be a bumpy ride in any case — though a lot bumpier if we do nothing. As I have emphasized already, much work needs to be done in terms of national and global economic and environmental policies (the kinds of responses discussed in Chapter 6); yet even if needed national and global monetary and energy reforms were to be enacted — and this would be no small accomplishment — we would still face decades of perilous environmental, economic, and social challenges.

Still, it's useful to contemplate a best-case outcome. Assuming we do everything right, what could we achieve? How might the world look as a result?

What the facts require is pretty clear: The best-case scenario we come up with, if it is to be realistic, must fit several non-negotiable criteria. The economy of the future will necessarily be steady-state, not requiring constant growth. It will be based on the use of renewable resources harvested at a rate slower than that of natural replenishment; and on the use of non-renewable resources at declining rates, with metals and minerals recycled and re-used wherever possible. Human population will have to achieve a level that can be supported by resources used this way, and that level is likely to be significantly lower than the current one.

But these criteria leave many details open to conjecture. What technologies could we develop and use under these conditions? Exactly what size of population would be sustainable? The two questions are related: the level of population that is sustainable will depend on what kind of technology we are able to develop and use. The supportable population size will also depend on how much we degrade soil, water, and climate *before* we achieve a condition of sustainability.

Some of the best recent writing on futurology is contained in John Michael Greer's *The Ecotechnic Future*. Greer, ever the historian, manages to be both realistic and hopeful:

> The generations that grow up in a world after industrialism will face many of the same kind of challenges that their ancestors did in the dark ages that followed other high civilizations. Some of those challenges must be confronted as they emerge. It may be possible, however, to counter or even forestall others by drawing on the resources of industrial civilization, to hand down valuable tools and insights to those who will need them. While Utopia is not an option, societies that are humane, cultured and sustainable are quite another matter. There have been plenty of them in the past; there can be many more in the future; and actions we can take today can help make that goal more accessible to the people of the ecotechnic age.[19]

When looking to the past for clues about how our lower-energy future might look and feel, it's hard to avoid becoming fixated on images from movies and television programs like *Little House on the Prairie* or *Brother Cadfael*. Some people find these depictions of a simpler life in bygone days attractive; others bristle at the thought that our descendants might have to do without computers, cell phones, and personal automobiles, busying themselves instead with plows and axes, herbs and chickens. Will the march of history pry our electronic toys from our cold, dead hands? How far back down the trail of complexity and technological sophistication might we have to retreat? Can we surrender cars, highways, and super-markets, but still keep cultural exchange, tolerance, and diversity, along with our hard-won scientific knowledge, advanced healthcare, and instant access to information?

That last question deserves considerable thought. During the past couple of centuries, energy consumption (and, more recently, debt) in-creased along with population and nasty environmental impacts. Social, cultural, and human benefits also proliferated. All three trends were closely related, with energy growth being the primary driver. In the de-cades ahead, available energy will decline. This will probably lead to de-clining population. It might or might not lead to declining environmental impacts (that depends on how we handle the transition: if we burn every last lump of coal, every last tree, and every last ton of tar sands, all in an effort to keep the lights on and the economy growing, then we might alter the energy decline rate at the cost of laying waste to the planet; on the other hand, if we reduce fossil fuel consumption proactively to protect the climate, we might preserve more of the biosphere at the cost of driving the energy curve down more sharply). But what about social and cultural benefits? Will they inevitably wither along with energy consumption?

This, I believe, will be one of the great questions and challenges of the coming century. As was argued in Chapter 6, if we focus on maximizing social and cultural benefits rather than on increasing GDP, we could actu-ally come out ahead in some respects—enjoying, for example, the sense of security that comes from knowing that the skills one learns today will still be relevant in twenty years, or from knowing that species one sees today will still be around for one's grandchildren to see as well.

Sometimes questions about what is possible can be answered with an experiment. In this case, the relevant experiment might consist of a com-munity trying to build a sustainable post-hydrocarbon economy while maximizing social and cultural payoffs. Dancing Rabbit Ecovillage is pre-cisely such a research project. Founded about 15 years ago by a small group of recent West Coast college graduates, Dancing Rabbit is an intentional community of about 50 people set amid the hills and prairies of rural northeastern Missouri. The stated goal of its residents is "to live ecologically sustainable and socially rewarding lives, and to share the skills and ideas behind that lifestyle."[20] Members of the community explore natural building, ecological food systems, and consensus decision-making; they have a vehicle co-op, a co-op healthcare fund, and guest space for visitors. By all accounts, the experiment is going well.

Having lived for years in intentional communities (back when I was younger and had more time on my hands), I know the challenges — and the rewards. It's not a lifestyle for everyone. It requires effort to start a community or to select and join one, and more effort to hash out all the rules and the conflicts that always come up in community life. Many communities fall by the wayside. Only a tiny fraction of the global population is involved in ecovillages or intentional communities, so we can't realistically expect this strategy to solve the world's problems. Nevertheless, thousands of such projects are making a go of it, and yielding useful knowledge and experience in the process.[21] The future may not look exactly like Dancing Rabbit, but it's easy to imagine worse outcomes and perhaps hard to think of much better ones.

Apart from intentional communities, there are thousands of organizations, companies, and individuals devoted to finding a path to sustainability — a way of life that will work not just for us, but for the seventh generation hence. These groups and individuals are active in nearly every city and town. You can usually locate them on the Internet with search words like Transition, permaculture, renewable energy, and appropriate technology.[22]

If there's anything to be known today about a positive future that may await us, it comes from the real-world efforts of people like these to solve practical problems — not from armchair forecasts based on current market trends.

Perspective

We are living through the fifth great turning in human history.[23]

The first was the harnessing of fire nearly two million years ago. Fire enabled us to stay warm in forbidding environments, cook our food (leading to profound changes not only in human culture but human physiology as well), and alter landscapes in our favor.[24]

The second was the development of language — likely a gradual process that began many tens of millennia ago, but an equally fateful one: it enabled humans to coordinate their actions over time and space, and it slowly altered the internal architecture of our brains. With language we told stories, and with those stories we wove religions, philosophies, and eventually scientific theories and computer programs.

The third turning point was the agricultural revolution 10,000 years ago. Seasonal surpluses of storable food enabled full-time division of labor (society became segmented into peasants, soldiers, accountants, merchants, and kings) as well as the emergence of cities and empires — which brought with them writing, mathematics, and money.

The industrial revolution, only about two centuries old, liberated the energies of fossil fuels, which replaced muscle power in production and transportation, thereby dramatically increasing the speed and scale of those processes. Fossil-fueled economic growth enabled human population to expand seven-fold and led to an explosion of scientific research, practical innovation, and trade. So much economic activity was now possible that a phenomenal expansion of credit was needed to connect potential producers with potential consumers.

Now we are participating in the turning from fossil fueled, debt- and growth-based industrial civilization toward a sustainable, renewable, steady-state society. While previous turnings entailed overall expansion (punctuated by periodic crises, wars, and collapses), this one will be characterized by an overall contraction of society until we are living within Earth's replenishable budget of renewable resources, while continually recycling most of the minerals and metals that we continue to use. No one who is alive today will be around to see the culmination of this fifth turning, and there is no way to know exactly what the end result will look like. The remainder of the current century will be a time of continual evolution and adaptation as we head, in fits and starts, toward that distant goal — which will itself be a dynamic rather than a static condition, in that human beings will still be evolving and society will still need to adapt continually to its changing environment.

There is no guarantee that the participants in this evolutionary and revolutionary transformation will view it as an extension of human progress, rather than as the ending of civilization as we have known it. Unless we completely fail to rise to the occasion, in which case the human project will simply cease, there will probably be elements of both collapse and renewal.

History suggests that it's hard to understand and deliberately navigate one of these great turnings as it is happening. Adam Smith lived through the early Industrial Revolution and laid the basis for the economic ideology that would shape the remainder of the industrial period. Yet there is no evidence in his writings that he understood the implications of the coal-based economy that was emerging around him. The thought that fossil fuels would utterly transform the world during the next two centuries evidently never entered his head. We can only speculate about his reaction if he could somehow be revived to witness today's massive global trade via fuel-powered ships, planes, trains, and trucks, or participate in a computer-mediated stock trade enabled by coal-fired electrons. When Smith imagined the economy of the future, he foresaw one comprised of shopkeepers, artisans, small factories, and trade via sailing ships, because those were his customary terms of reference.

We're at a similar juncture today. Before us lies a future that will necessarily be very different from the one that our political leaders encourage us to envision. The only mental tools we have with which to imagine the possibilities that await us are ones honed in the past era of growth, extraction, and combustion. As a result, we can't hope to have a very clear picture of what life will or even could be like for grandchildren of the children now being born.

What we *can* hope to do is to make sure they have a future — that they will have even the possibility of existing and making their own contributions to our species' unfolding story. In order for future generations to enjoy the barest of chances at life we must avoid the monetary-financial wall in our path, or ensure that the impact is minimal. And we must set a course toward sustainability and away from collision with Earth's environmental limits. All of this will require us to question what we think we know, to leave our comfort zones far behind, and to engage in hard, challenging work.

We will be tempted to waste time fussing over aspects of our current way of life that may not be salvageable (including many of the goods we associate with economic growth). We will be tempted also to waste time apportioning blame for the failure of our existing economic and industrial systems, and venting anger over the greed and stupidity that stand in the way of building a new economy. None of this will help. The only efforts that will aid in the long run are those that contribute, in some tangible way, to the realization of a pattern of human settlement that is culturally and psychologically rewarding, and that supports rather than undermines the integrity of Earth's living skin, our only home.

Notes

Introduction

1. Donella Meadows, Jorgen Randers, Dennis Meadows, and William Behrens III, *Limits to Growth*. (New York, Signet, 1972)
2. Ugo Bardi, "Cassandra's Curse: How 'The Limits to Growth' Was Demonized," *The Oil Drum*, posted March 9, 2008, europe.theoildrum.com/node/3551.
3. Graham Turner, "A Comparison of the Limits to Growth with Thirty Years of Reality," CSIRO Sustainable Ecosystems, June 2008.
4. Donella Meadows, Jorgen Randers, and Dennis Meadows, *Limits to Growth: The 30-Year Update* (White River Junction, VT: Chelsea Green, 2004); Sam Foucher, "Ecological Footprint, Energy Consumption, and the Looming Collapse," posted May 16, 2007, theoildrum.com/node/2534.
5. According to the Federal Reserve, total commercial debt in the US is about $36 trillion. If the average interest on this debt is 3 percent, probably a very low estimate, then we owe in interest about $1 trillion per year. That's over 6 percent of GNP. If GNP grows at 3 percent, then recipients of interest get all new income plus 3 percent of the old income, until eventually they end up with everything. It takes a lot of growth just to enable repayment of interest.
6. Growth had actually begun in Britain around 1500 as a result of colonialism (an early form of globalization), intensification of agriculture, division of labor, and the adoption of more wind and water power. However, it proceeded at a comparatively slow pace until the adoption of coal-fired machinery in the late 18th century. During the 20th century, global growth averaged about 3 percent per annum. See Stephen Broadberry et al., "British Economic Growth, 1270–1870," University of Warwick, UK, published online December 6, 2010.
7. "GDP United States (Recent History)," *Data360.org*, data360.org/dsg.aspx?Data_Set_Group_Id=353&page=3&count=100.
8. "'Great Recession' Over, Research Group Says," *msnbc.msn.com*, posted September 20, 2010.
9. Carmen M. Reinhart and Kenneth S. Rogoff, "The Aftermath of Financial Crises," (presented at the meeting of the American Economic Association, San Francisco, CA, January 3, 2009).
10. In philosophy this is called the problem of induction. One cannot infer that a series of events will happen in the future just as they have in past. For example, if all the swans I've ever seen are white, I may conclude that all swans are white. However, consistently seeing white swans does not prove that all swans are white; it only adds to a series of observations. The inductive conclusion that all swans are white is proven wrong with the discovery of even one black swan. Nassem Talib's book *Black Swan: The Impact of the Highly Improbable* explores this topic in-depth (New York: Random House, 2007).

11. In inflation-adjusted dollars. The amount of energy, in British Thermal Units, required to produce a dollar of GDP dropped from close to 20,000 Btu per dollar in 1949 to 8,500 Btu in 2008. Source: US Department of Energy.

12. Praveen Ganta, "US Economic Energy Efficiency: 1950–2008," *Seeking Alpha*, posted January 10,2010.

13. See Alan Hastings, *Population Biology* (New York: Springer, 1996), and William Catton, *Overshoot: The Ecological Basis of Revolutionary Change* (Urbana: University of Illinois Press, 1982).

14. Colin Campbell, "Colin Campbell Predicts Credit Crunch Due to Peak Oil 2005," YouTube video, posted October 16, 2008, youtube.com/watch?v=lDNMjV6sumQ&feature=related.

15. The Peak Oil scenario is laid out in more detail in my book *The Party's Over: Oil, War and the Fate of Industrial Societies* (Gabriola Island, BC: New Society, 2003).

16. Richard Heinberg, "Temporary Recession or the End of Growth?" *Energy Bulletin*, posted August 6, 2009, energybulletin.net/node/49798.

17. Gail the Actuary, "What Peaked at the Same Time as Oil Prices? Lots of Things," *The Oil Drum*, posted October 9, 2009, theoildrum.com/node/5853.

18. Joe Keohane, "The Guy Who Called the Big One? Don't Listen to Him: Inside the Paradox of Forecasting," *boston.com*, posted January 9, 2011.

19. David Murphy, "Lucky Economists, Unlucky Scientists?" *The Oil Drum*, posted January 14, 2011, theoildrum.com/node/7343.

Chapter One

1. See Marvin Harris and Orna Johnson, *Cultural Anthropology* (Boston: Allyn & Bacon, 2006), pp. 98–107. See also Marcel Mauss, *The Gift: The Form and Reason for Exchange in Archaic Societies*, transl. I. Cunnison (New York: Norton, 2000), and David Graeber, *Toward an Anthropology of Value* (New York: Palgrave, 2001).

2. Elman R. Service, *The Hunters* (New York: Prentice Hall, 1966), pp. 14–21.

3. Fernand Braudel, *The Structures of Everyday Life* (Berkeley: University of California Press, 1992), p. 436.

4. This story is told at greater length in, for example, Jack Weatherford, *The History of Money* (New York: Three Rivers Press, 1998).

5. Mark Anielski, "Fertile obfuscation: Making money whilst eroding living capital" (Pembina Institute, 2000); Michael Rowbathom, *The Grip of Death: A Study of Modern Money, Debt Slavery, and Destructive Economics* (New York: Jon Carpenter, 1998).

6. Tim Parks, *Medici Money* (New York: Norton, 2005), p. 13.

7. The history of economics is told entertainingly and in more detail in lectures by Australian economist Steve Keen, notes for which are available at debunkingeconomics.com/Lectures/Index.htm#HET.

8. See Steve Keen, *Debunking Economics: The Naked Emperor of the Social Sciences* (London: Zed Books, 2002).

9. John Stuart Mill. Quoted in Brian Czech, *Shoveling Fuel for a Runaway Train* (Berkeley and Los Angeles: University of California Press, 2000), p. 93.

10. This paragraph is based on historical analysis and insights from Herman Daly and John Cobb. See Herman Daly and John Cobb, *For the Common Good* (Boston: Beacon Press, 1994), pp. 97–120.

11. For a further discussion of this point see Daly and Cobb, *op. cit.*
12. A periodically updated discussion of the business cycle from various points of view is available at Wikipedia.
13. Though not according to economists who subscribe to the Elliott Wave theory. See A. J. Frost and Robert Prechter, *Elliott Wave Principle: Key to Market Behavior* (Gainesville, GA: New Classics Library, 2006).
14. This appears to be the opposite of what is happening now — money is tight but interest rates are low. The current situation exists because the Federal Reserve is deliberately keeping interest rates low to stimulate economic activity.
15. See Bruce Jansson, *The Sixteen-Trillion-Dollar Mistake* (New York: Columbia University Press, 2002).
16. Perhaps the biggest difference between the two forms of investment shows up in demand curves. Speculative demand increases as prices rise. Investment demand decreases as prices of the capital being invested in rise. The notion of market equilibrium is based on negative feedback loops, while speculation creates positive feedback loops.
17. Charles McKay, *Extraordinary Popular Delusions and the Madness of Crowds* (1841, various editions).
18. It is useful to distinguish among hedge, speculative, and ponzi finance. Hedge: investors have money to repay both capital and interest on the loan. Speculative: investors can only pay interest and count on rising values or new loans to pay capital (example: subprime, interest-only loans). Ponzi: investors cannot even pay interest, and count on rising values to repay both capital and interest (example: subprime loans in which interest is added to capital).
19. Investments that systematically generate returns higher than economic growth are impossible to sustain in the long run, as investors would eventually come to own all wealth.

Chapter Two

1. Andrew Ross Sorkin, *Too Big to Fail* (New York: Viking, 2009); Bethany McLean and Joe Nocera, *All the Devils Are Here: The Hidden History of the Financial Crisis* (New York: Portfolio, 2010); Charles Ferguson, "Inside Job", documentary movie, Sony Pictures Classics, 2010.
2. William Greider, "The Education of David Stockman," *Atlantic Magazine*, December, 1981.
3. Both the president and Congress pledged to put in place new regulations that would make the recurrence of such a fiasco impossible. The result was a mild rewrite of laws; according to Christine Harper of Bloomberg, "Lawmakers spurned changes that would wall off deposit-taking banks from riskier trading. They declined to limit the size of lenders or ban any form of derivatives. Higher capital and liquidity requirements agreed to by regulators worldwide have been delayed for years to aid economic recovery." Christine Harper, "Out of Lehman's Ashes Wall Street Gets Most of What It Wants," *Bloomberg Businessweek*, January 5, 2011.
4. Financial Crisis Inquiry Commission, preliminary report, February 2011, fcic.gov.
5. The term "Giant Pool of Money" became the title of an award-winning episode of the NPR radio show *This American Life* which originally aired on May 9, 2008.

6. James B. Stewart, "Eight Days: the battle to save the American financial system", *The New Yorker* magazine, September 21, 2009.

7. Financial Crisis Inquiry Commission, preliminary report, February 2011, fcic.gov.

8. Report of the Financial Crisis Inquiry Commission, fcic.law.stanford.edu.

9. "The global housing boom." *The Economist*, June 16 2005.

10. See for example Harry S. Dent, *The Great Depression Ahead* (New York: Free Press, 2009), pp. 132–133.

11. Dowell Myers and SungHo Ryu, "Aging baby boomers and the generational housing bubble: Foresight and mitigation of an epic transition" (2007). *Journal of American Planning Association*, 74:1, pp. 17–33.

12. Michael Kumhof and Romain Rancière, "Inequality, Leverage and Crises," IMF Working Paper (2010), imf.org/external/pubs/ft/wp/2010/wp10268.pdf.

13. US Department of the Treasury, *TreasuryDirect*, "Historical Debt Outstanding, 2000–2010," treasurydirect.gov/govt/reports/pd/pd.htm.

14. Congressional Budget Office, *Report to the Joint Committee on Taxation*, March 5, 2010.

15. Congressional Budget Office, "Monthly Budget Review," October 7, 2010.

16. James Galbraith, "Why We Don't Need to Pay Down the National Debt." *The Atlantic*, 9/21/2010theatlantic.com/business/archive/2010/09/why-we-dont-need-to-pay-down-the-national-debt/63273.

17. For a perspective on why US government debt may not face limits anytime soon, as long as the economy returns to growth, see James K. Galbraith, "Casting Light on 'The Moment of Truth,'" *The Huffington Post*, posted December 3, 2010.

18. Carmen M. Reinhart and Kenneth Rogoff, *This Time Is Different: Eight Centuries of Financial Folly* (New Jersey: Princeton University Press, 2009).

19. See Paul Krugman, "The burden of debt," New Y *Times*. 4/28/2009. krugman.blogs. nytimes.com/2009/08/28/the-burden-of-debt/; Daniel Berger, "The Deficit: Size Doesn't Matter" (2009). Roosevelt Institute. rooseveltinstitute.org/new-roosevelt/deficit-size-doesn-t-matter.

20. Michael J. Panzner, *Financial Armageddon* (New York: Kaplan Publishing, 2008), p. 45.

21. Ellen Hodgson Brown, *The Web of Debt* (Boxboro MA: Third Millenium Press, 2010), p. 197.

22. McKinsey & Co., "Debt and Deleveraging: The Global Credit Bubble and Its Economic Consequences" (2010), mckinsey.com/mgi/publications/debt_and_deleveraging/index.asp.

23. Herman Daly, *Beyond Growth: The Economics of Sustainable Development* (Boston: Beacon, 1997).

24. CNN Money.com, "Bailout Tracker," money.cnn.com/news/storysupplement/economy/bailouttracker/index.html.

25. Christian Broda and Jonathan Parker, "The impact of 2008 tax rebates on consumer spending: preliminary evidence" (6/29/2008). insight.kellogg.northwestern.edu/index.php/Kellogg/article/the_impact_of_the_2008_tax_rebates_on_consumer_spending.

26. See, for example, heritage.org/research/reports/2010/01/why-government-spending-does-not-stimulate-economic-growth-answering-the-critics.

27. Pam Martens, "The Tax-Payers' Tab: a Cool $9 Trillion and Then Some," *Counterpunch*, posted December 20, 2010.

28. For example, see Vox Day, *The Return of the Great Depression* (Los Angeles: WND Books, 2009).

29. "QE2: More Effective Than QE1," *International Business Times*, posted November 7, 2010.

30. Henny Sender, "Spectre of Deflation Kills the Mood at Jackson Hole," *Financial Times*, September 13, 2010.

31. L. Randall Wray, "Why Mortgage-Backed Securities Aren't (Backed by Securities): How MERS Toasted the Banks," *The Huffington Post*, posted December 30, 2010; Yasha Levine, "Dude, Where's My Mortgage? How an Obscure Outfit Called MERS Is Subverting Our Entire System of Property Rights," AlterNet.org, posted December 16, 2010.

32. One of the most prominent inflationists is investor Jim Rogers; see, for example, "Jim Rogers Says Hyperinflation Will Consume Us," allthingsjimrogers.com/2009/03/12/jim-rogers-says-hyperinflation-will-consume-us.

33. See, for example, John Williams, shadowstats.com/article/hyperinflation-2010.

34. See, for example, Nicole Foss, theautomaticearth.blogspot.com; also Harry S. Dent, Jr., *The Great Depression Ahead* (New York: Free Press, 2009).

35. See, for example, Nicole Foss ("Stoneleigh"), "The Unbearable Mightiness of Deflation." theautomaticearth.blogspot.com/2009/07/july-5-2009-unbearable-mightiness-of.html.

36. This is the position of Nicole Foss; see "Stoneleigh Takes on John Williams," theautomaticearth.blogspot.com/2010/08/august-8-2010-stoneleigh-takes-on-john.html.

Chapter Three

1. There are now many books on Peak Oil, among which some of the best are: Kenneth S. Deffeyes, *Beyond Oil: The View from Hubbert's Peak* (New York: Hill and Wang, 2005); Matthew R. Simmons, *Twilight in the Desert: The Coming Saudi Oil Shock and the World Economy* (New York: Wiley, 2005); Paul Roberts, *The End of Oil: On the Edge of a Perilous New World* (New York: Houghton Mifflin, 2004); Richard Heinberg, *The Party's Over: Oil, War and the Fate of Industrial Societies* (Gabriola Island, BC: New Society, 2005). Two websites offer daily updated links to articles relevant to Peak Oil: EnergyBulletin.net and TheOilDrum.com. The Association for the Study of Peak Oil and Gas, with chapters in many countries, convenes annual conferences for the discussion of new data relevant to oil depletion and our energy future.

2. "...[In 2035] Global oil production reaches 96 mb/d, the balance of 3 mb/d coming from processing gains. Crude oil output reaches an undulating plateau of around 68–69 mb/d by 2020, but never regains its all-time peak of 70 mb/d reached in 2006, while production of natural gas liquids (NGLs) and unconventional oil grows strongly." International Energy Agency, "Executive Summary," *World Energy Outlook 2010* (Paris: OECD/IEA, 2010), p. 6; Also see the graph "World oil production by type in the New Policies Scenario," IEA, "Key Graphs," *World Energy Outlook 2010*, worldenergyoutlook.org/docs/weo2010/key_graphs.pdf.

3. Kjell Aleklett, "Spin Slips Off Production Numbers—World Energy Outlook 2010 Is a Cry For Help," *Aleklett's Energy Post*, posted November 11, 2010.

4. Jeremy Gilbert, "No We Can't: Uncertainty, Technology, and Risk," presented at the ASPO-USA 2010 Peak Oil Conference, Washington, DC, October 9, 2010.

5. Jessica Bachman, "Special Report: Oil and Ice: Worse Than the Gulf Spill?" *Reuters .com*, posted November 8, 2010.

6. Charles Maxwell, quoted in Wallace Forbes, "Bracing For Peak Oil Production by Decade's End," *Forbes.com*, posted September 13, 2010; Eoin O'Carroll, "Pickens: Oil Production Has Peaked," *The Christian Science Monitor*, posted June 18, 2008.

7. Clint Smith, "New Zealand Parliament Peak Oil Report: The Next Oil Shock?" *Energy Bulletin*, posted October 1, 2010, energybulletin.net/stories/2010-10-14/next-oil-shock; Stefan Schultz, "Military Study Warns of a Potentially Drastic Oil Crisis," *Spiegel Online*, posted September 1, 2010; UK Industry study; US Joint Forces Command, *The Joint Operating Environment 2010* (Suffolk, VA: USJFCOM, 2010).

8. See, for example, peakoil.net/headline-news/toyota-we-must-address-the-inevitability -of-peak-oil.

9. Ben German, "The Other Peak Oil: Demand From Developed World Falling," *Scientific American*. October 13, 2009.

10. Robert L. Hirsch, Roger Bezdek, and Robert Wendling, "Peaking of World Oil Production: Impacts, Mitigation, & Risk Management," a report for the US Department of Energy, National Energy Technology Laboratory, February 2005; Stefan Schultz, "Military Study Warns of a Potentially Drastic Oil Crisis," *Spiegel Online*, posted September 1, 2010.

11. Michael Klare, *Blood and Oil: The Dangers and Consequences of America's Growing Petroleum Dependency* (New York: Henry Holt and Co., 2004); Michael Klare, *Rising Powers, Shrinking Planet: The New Geopolitics of Energy* (New York: Henry Holt and Co., 2008).

12. Jeffrey Brown and Sam Foucher, "Peak Oil Versus Peak Exports," *Energy Bulletin*, posted October 18, 2010, energybulletin.net/stories/2010-10-18/peak-oil-versus-peak -exports; Gail the Actuary, "Verifying the Export Land Model — A Different Approach," *The Oil Drum*, posted October 1, 2010, theoildrum.com/node/7007#more.

13. Thanks to Nate Hagens for this insight.

14. Buttonwood, "Engine Trouble: A Rise in the Cost of Extracting Energy Will Hit Productivity," *The Economist*, October 1, 2010.

15. Tadeusz W. Patzek and Gregory D. Croft, "A Global Coal Production Forecast with Multi-Hubbert Cycle Analysis," *Energy* 35 (2010), pp. 3109–3122; S. H. Mohr and G. M. Evans, "Forecasting Coal Production Until 2100," *Fuel* 88 (2009), pp. 2059–2067.

16. Richard Heinberg and David Fridley, "The End of Cheap Coal," *Nature* 468 (November 18, 2010), pp. 367–369.

17. Recent US Geological Survey assessments of some of the most important mining regions show rapid depletion of accessible reserves. US Geological Survey, *Coal Reserves of the Matewan Quadrangle, Kentucky — A Coal Recoverability Study*, US Bureau of Mines Circular 9355, pubs.usgs.gov/usbmic/ic-9355/C9355.htm; James Luppens et al., *Assessment of Coal Geology, Resources, and Reserves in the Gillette Coalfield, Powder River Basin, Wyoming*, USGS Open-File Report 2008–1202, 2008, pubs.usgs.gov/of /2008/1202/.

18. Zaipu Tao and Mingyu Li, "What Is the Limit of Chinese Coal Supplies — A STELLA Model of Hubbert Peak," *Energy Policy* 35, no.6 (June, 2007), pp. 3145–3154; WorleyParsons et al., *Strategic Analysis of the Global Status of Carbon Capture and Storage* (Global CCS Institute, 2009).

19. Christopher Schenk and Richard Pollastro, *Natural Gas Production in the United*

States, US Geological Survey Fact Sheet FS-0113-01, January 2002, pubs.usgs.gov/fs/fs -0113-01/.

20. Gail the Actuary, "Arthur Berman Talks About Shale Gas," *The Oil Drum*, posted July 28, 2010, theoildrum.com/node/6785.

21. A report by David Hughes on the future of unconventional natural gas will be published in May 2011 by Post Carbon Institute.

22. Richard Heinberg, *Searching for a Miracle: Net Energy Limits and the Fate of Industrial Societies* (International Forum on Globalization and Post Carbon Institute, 2009).

23. This conclusion is supported by P. Moriarty and D. Honnery, "What Energy Levels Can the Earth Sustain?" *Energy Policy* 37, no. 7 (July, 2009), pp. 2469–2474. Abstract: "Several official reports on future global primary energy production and use develop scenarios which suggest that the high energy growth rates of the 20th century will continue unabated until 2050 and even beyond. In this paper we examine whether any combination of fossil, nuclear, and renewable energy sources can deliver such levels of primary energy—around 1000 EJ in 2050. We find that too much emphasis has been placed on whether or not reserves in the case of fossil and nuclear energy, or technical potential in the case of renewable energy, can support the levels of energy use forecast. In contrast, our analysis stresses the crucial importance of the interaction of technical potentials for annual production with environmental factors, social, political, and economic concerns and limited time frames for implementation, in heavily constraining the real energy options for the future. Together, these constraints suggest that future energy consumption will be significantly lower than the present level."

24. Heinberg, *Searching for a Miracle*.

25. Heinberg, *Searching for a Miracle*, chapter 3.

26. This conclusion is largely supported by James H. Brown et al., "Energetic Limits to Growth," *Bioscience* 61, no.1 (January 2011), pp. 19–26: "The bottom line is that an enormous increase in energy supply will be required to meet the demands of projected population growth and lift the developing world out of poverty without jeopardizing current standards of living in the most developed countries. And the possibilities for substantially increasing energy supplies are highly uncertain. Moreover, the nonlinear, complex nature of the global economy raises the possibility that energy shortages might trigger massive socioeconomic disruption."

27. Nate Hagens, "Applying Time to Energy Analysis," *The Oil Drum*, posted December 13, 2010, theoildrum.com/node/7147#more.

28. Charles A. S. Hall, "Provisional Results from EROI Assessments," *The Oil Drum*, posted April 8, 2008, theoildrum.com/node/3810.

29. "Drowning in Oil," *The Economist*, March 4, 1999.

30. International Energy Agency, *World Energy Outlook 1998* (Paris: IEA Publications, 1998), p. 78.

31. Chris Nelder, "World Will Soon Face an Oil Supply Crunch," *Energy & Capital*, posted September 18, 2009.

32. Richard Heinberg, "Goldilocks and the Three Fuels," *Reuters*, posted February 18, 2010.

33. James Hamilton, *Causes and Consequences of the Oil Shock of 2007*, Brookings Papers on Economic Activity, March 2009.

34. Joe Cortright, "Driven to the Brink: How the Gas Price Spike Popped the Housing Bubble and Devalued the Suburbs," White Paper, *CEOs for Cities*, 2008, ceosforcities .org.

35. Jad Mouawad, "For OPEC, Current Oil Price Is Just Right," *The New York Times*, September 9, 2009.

36. Jon Talton, "With Oil Prices Around $90, Recovery is Over a Barrel," *The Seattle Times*, December 9, 2010; David Murphy, "Further Evidence of the Influence of Energy on the US Economy," *The Oil Drum*, posted April 16, 2009, netenergy.theoildrum.com /node/5304; Jeff Rubin, "We Have Run Out of Oil We Can Afford to Burn," *The Globe and Mail*, October 6, 2010; "Oil Price is Risk to Economic Recovery, Says IEA," *BBC News*, posted January 5, 2011; Derek Thompson, "How Oil Could Kill the Recovery," *The Atlantic*, posted January 6, 2011.

37. Cameron Leckie, "Economic Growth: A Zero Sum Game," *On Line Opinion*, posted November 25, 2010.

38. Carey W. King, "Energy Intensity Ratios as Net Energy Measures of United States Energy Production and Expenditures," *Environmental Research Letters* 5 (October to December 2010).

39. Recent reports warn that groundwater is depleting at increasing rates "Groundwater Depletion Rate Accelerating Worldwide," *Science Daily*, posted September 23, 2010; and that the world's rivers are in a "crisis state." World's Rivers in 'Crisis State', Report Finds," *Science Daily*, posted October 1, 2010. The situation in many areas is grim and worsening. Robert F. Worth, "Earth is Parched Where Syrian Farms Thrived," *The New York Times*, October 13, 2010. For an excellent overview of the situation, see Sandra Postel, "Water: Adapting to a New Normal" in *The Post Carbon Reader*, Richard Heinberg and Daniel Lerch, eds. (Healdsburg, CA: Watershed Media and the Post Carbon Institute, 2010); see also Peter H. Gleick and Meena Palaniappan, "Peak Water Limits to Freshwater Withdrawal and Use," *Proceedings of the National Academy of Science* 107, no.25 (June 22, 2010), pp. 11155–11162.

40. United Nations Environment Program, *Global Environment Outlook 4* (Malta: Progress Press Ltd., 2007).

41. Yoshihide Wada et al., "Global Depletion of Groundwater Resources," *Geophysical Research Letters* 37 (October 26, 2010).

42. Felicity Barringer, "Rising Seas and the Groundwater Equation," Green: A Blog About Energy and the Environment, *The New York Times*, posted November 2, 2010.

43. Tim P. Barnett and David W. Pierce, "When Will Lake Mead Go Dry?" *Water Resources Research* 44 (March 29, 2008).

44. National Energy Technology Laboratory, Innovations for Existing Plants Program, "Water-Energy Interface," netl.doe.gov/technologies/coalpower/ewr/water/power-gen .html.

45. "The United Nations Intergovernmental Panel on Climate Change famously predicted [the Himalayan glaciers] could disappear as soon as 2035. It turns out that guesstimate was based on misquoting a researcher in a 1999 news article — not a result from any kind of peer-reviewed scientific study. The incident reflects a breakdown in the IPCC process but it doesn't undercut the reality that glacier loss, particularly in what are technically tropical regions such as the Andes and Himalayas, continues to accelerate in the 21st century. Though they likely won't disappear entirely for centuries, losing the glaciers will eventually be bad news for the billions around the world who rely on meltwater to survive." David Biello, "How Fast Are the Himalayan Glaciers Melting?" *Scientific American* podcast, posted January 21, 2010, scientificamerican.com/podcast /episode.cfm?id=how-fast-are-himalayan-glaciers-mel-10-01-21. Actual melt rates

are a matter of ongoing study, but there is general agreement that, on the whole, the glaciers are retreating rapidly. In the Indian Himalaya, the Chhota Shigri Glacier has retreated 12 percent in the past 13 years and the iconic Gangotri Glacier, where the River Ganga originates, has retreated 12 percent in the past 16 years. See Richard S. Williams, Jr., and Jane G. Ferrigno, eds., *Glaciers of Asia*, US Geological Survey Professional Paper 1386–F (Washington, DC: US GPO, 2010), online at pubs.usgs.gov/pp/p1386f/.

46. "Desperate Water Shortage in Somaliland," *Inside Somalia*, posted August 4, 2009.

47. Nancy L. Barber, "Summary of Estimated Water Use in the United States in 2005," US Geological Survey, 2009, water.usgs.gov/watuse/.

48. Kevin F. Dennehy, *High Plains Regional Groundwater Study*, U. S. Geological Survey Fact Sheet FS-091-00, 2000, co.water.usgs.gov/nawqa/hpgw/PUBS.html.

49. Paul D. Ryder, "High Plains Aquifer," in *Groundwater Atlas of the United States: Oklahoma, Texas*, US Geological Survey publication HA 730-E, 1996, pubs.usgs.gov/ha/ha730/index.html.

50. Barber, "Summary of Estimated Water Use in the United States in 2005."

51. "Industrial-Agricultural Water End-Use Efficiency," *California Energy Commission* website, energy.ca.gov/research/iaw/industry/water.html.

52. "Membrane Desalination Power Usage Put in Perspective," *American Membrane Technology Association*, April 2009, amtaorg.com/amta_media/pdfs/7_Membrane DesalinationPowerUsagePutInPerspective.pdf.

53. Jeremy Miller, "California Drought is No Problem for Kern County Oil Producers," *Circle of Blue*, posted August 24, 2010.

54. Barber, "Summary of Estimated Water Use in the United States in 2005."

55. Peter Boaz and Matthew O. Berger, "Rising Energy Demand Hits Water Scarcity 'Choke Point'," IPSNews.net, posted September 22, 2010, ipsnews.net/news.asp?idnews =52939.

56. "Ethiopia and Egypt Dispute the Nile," *BBC News*, posted February 24, 2005.

57. Alistair Lyon, "Arab World to Face Severe Water Scarcity By 2015," *Ottawa Citizen*, November 4, 2010.

58. In his book, *Dirt*, David Montgomery makes a powerful case that soil erosion was a major cause of the Roman economy's decline. David Montgomery, *Dirt: The Erosion of Civilizations* (Berkeley and Los Angeles: University of California Press, 2007).

59. T. Beach et al., "Impacts of the Ancient Maya on Soils and Soil Erosion in the Central Maya Lowlands," *Catena* 65, no.2 (February 28, 2006), pp. 166–178.

60. Richard Heinberg and Michael Blomford, *The Food and Farming Transition*, Post Carbon Institute, 2009, available online postcarbon.org/report/41306-the-food-and -farming-transition-toward.

61. Jonathan Foley, "The *Other* Inconvenient Truth: The Crisis in Global Land Use," *environment 360*, posted October 5, 2009.

62. "World Fertilizer Consumption," spreadsheet for "Food and Agriculture," *Earth Policy Institute* Data Center, posted January 12, 2011, earth-policy.org/data_center/C24; Robert J. Diaz et al., "Spreading Dead Zones and Consequences for Marine Ecosystems," *Science* 321, no.926 (2008).

63. David Tilman et al., "Agricultural Sustainability and Intensive Production Practices," *Nature* 418 (August 8, 2002), pp. 671–677.

64. Scott Kilman and Liam Pleven, "Harvest Shocker Rattles Wall Street," *The Wall Street Journal*, October 9, 2010.

65. Food and Agriculture Organization of the United Nations, "Protect and Produce: Restoring the Land," in *Dimensions of Need — An Atlas of Food and Agriculture* (Rome: FAO, 1995); Leo Horrigan, Robert S. Lawrence, and Polly Walker, "How Sustainable Agriculture Can Address the Environmental and Human Health Harms of Industrial Agriculture," *Environmental Health Perspectives* 110, no.5 (May, 2002).

66. Patrick Déry and Bart Anderson, "Peak Phosphorus," *The Oil Drum*, posted August 17, 2007, theoildrum.com/node/2882.

67. Patrick Déry, *Pérenniser l'agriculture*, Mémoire pour la Commission Sur l'Avenir de l'Agriculture du Québec, GREB, April 2007.

68. Juliet Eilperin, "World's Fish Supply Running Out, Researchers Warn," *The Washington Post*, November 3, 2006.

69. Corinne Podger, "Depleting Fish Stocks," *BBC World Service*, posted August 29, 2000.

70. Julian Cribb, *The Coming Famine: The Global Food Crisis and What We Can Do to Avoid It* (Berkeley and Los Angeles: University of California Press, 2010).

71. R. D. Howard, J. A. DeWoody, and W. M. Muir, "Transgenic Male Mating Advantage Provides Opportunity for Trojan Gene Effect in Fish," *Proceedings of the National Academy of Science* 101, no.9 (February 19, 2004), pp. 2934–2938.

72. Darrel Good, "Corn and Soybean Supplies to Remain Tight for Another Year?" University of Illinois, *ACES News*, posted Janury 24, 2011.

73. Lester Brown, "The Great Food Crisis of 2011," Foreign Policy, posted January 10, 2011.

74. *A Rock and a Hard Place: Peak Phosphorus and the Threat to our Food Security* (Bristol UK: Soil Association, 2010).

75. Recent analysis suggests that we may hit "peak" phosphate as early as 2033, after which supplies will become increasingly scarce and more expensive. Dana Cordell, Jan-Olof Drangert, and Stuart White, "The Story of Phosphorus: Global Food Security and Food for Thought," *Global Environmental Change* 19, no.2 (May 2009), pp. 292–305.

76. "New Threat to Global Food Security as Phosphate Supplies Become Increasingly Scarce," *Soil Association*, posted November 29, 2010.

77. Brian Palmer, "Has the Earth Run Out of Any Natural Resources?," *Slate*, posted October 10, 2010.

78. Craig Clugston, "Increasing Global Nonrenewable Natural Resource Scarcity — An Analysis," *Energy Bulletin*, posted April 6, 2010.

79. James Kanter, "Europe Sounds Alarm on Minerals Shortage," Green: A Blog About Energy and the Environment, *The New York Times*, posted June 16, 2010; Damien Gurco et al., *Peak Minerals in Australia: A Review of Changing Impacts and Benefits*, (Australia Commonwealth Scientific Industrial Research Organization, March 2010); David Cohen, "Earth's Natural Wealth: An Audit," *New Scientist* 2605 (May 23, 2007), pp. 34–41; Michael Moyer, "How Much Is Left?," *Scientific American* online, multimedia presentation, posted August 24, 2010.

80. Thomas D. Kelly and Grecia R. Matos, "Historical Statistics for Mineral and Material Commodities in the United States," United States Geological Survey, Data Series 140, 2010, minerals.usgs.gov/ds/2005/140/.

81. Peter Goodchild, "Depletion of Key Resources: Facts at Your Fingertips," *Culture Change*, posted January 27, 2010.

82. R. B. Gordon, M. Bertram, and T. E. Graedel, "Metal Stocks and Sustainability," *Proceedings of the National Academy of Sciences*, 103, no.5 (January 31, 2006), p. 1209.

83. Walter Youngquist, *Geodestinies: The Inevitable Control of Earth Resources Over Nations and Individuals* (Portland OR: National Book Company, 1997).

84. Julian Phillips, *Goldforecaster*, goldforecaster.com.

85. Office of Policy and International Affairs, *Critical Materials Strategy*, US Department of Energy, December, 2010.

86. David Cohen, "Earth's Natural Wealth: An Audit," *New Scientist* 2605 (May 23, 2007).

87. Emma Woollacott, "Shortage of Alternative Energy Minerals Will Trigger Trade Wars," *TGDaily*, posted November 1, 2010.

88. Cohen, "Earth's Natural Wealth," *New Scientist*.

89. "State Palladium Stockpile Nears Depletion," *The Moscow Times*, October 11, 2010.

90. Thomas Seltmann, "Nuclear Power: The Beginning of the End," *Sun & Wind Energy* (Energy Watch Group, November 2009).

91. "Peak Everything," *Wired.com*, wired.com/inspiredbyyou/2010/10/peak-everything /?ibypid=13.

92. Aeldric, "The Networking of Resource Production: Do the Networks Give Us Warnings When They Are About to Fail?," *The Oil Drum*, posted September 22, 2010, anz .theoildrum.com/node/6974#more.

93. Paul Crutzen, "The 'Anthropocene'," in *Earth System Science in the Anthropocene: Emerging Issues and Problems*, Eckart Ehlers and Thomas Krafft, eds. (New York: Springer, 2006).

94. Joseph A. Tainter, *The Collapse of Complex Societies*, New Studies in Archaeology, (Cambridge UK: Cambridge University Press, 1988).

95. Graeme Wearden, "BP oil spill costs to hit $40bn," *Guardian.co.uk*, posted November 2, 2010.

96. "Pakistan Flood-Related Losses to Reach 43 Billion Dollars," *Earth Times*, posted September 1, 2010; "Report: Wildfires, Drought Costing Russia $15 Billion," *Voice of America News.com*, posted August 10, 2010.

97. The following source lists total costs of natural disasters for the year at $109 billion, but this does not include the $40 billion Deepwater Horizon spill. Pat Speer, "Natural Disasters Cost $109 Billion in 2010," *Insurance Networking News*, posted January 24, 2011.

98. Subhankar Banarjee, "5 Mining Projects That Could Devastate the Entire Planet," AlterNet, posted November 16, 2010.

99. John Carey, "Calculating the True Cost of Global Climate Change," *environment360*, posted January 6, 2011.

100. "Stern Review: Unfavorable Critical Response," *Wikipedia*, accessed January, 2011.

101. Two other strategies include capturing and sequestering the carbon from fossil fuel combustion, and capturing and sequestering atmospheric carbon. Currently human efforts along these lines (ignoring, for the moment, the natural ongoing carbon capturing processes in soils, forests, and oceans) are making only an insignificant difference in the rate of growth in atmospheric greenhouse gases.

102. Heinberg, *Searching for a Miracle*.

103. Juliette Jowit, "One in Five Plant Species Face Extinction," *The Guardian*, September 29, 2010.

104. Frank Stephenson, "A Tale of Taxol," Florida State University, *Research in Review*, rinr .fsu.edu/fall2002/taxol.html.

105. Jonathan Watts, "Biodiversity Loss Seen As Greater Financial Risk Than Terrorism," *The Guardian*, October 27, 2010.

106. Richard Black, "Plankton Decline Across Oceans As Waters Warm," *BBC News*, posted July 28, 2010.

107. Peter Tatchell, "The Oxygen Crisis," *The Guardian*, August 13, 2008.

108. Matt Chorley, "$5,000,000,000,000: The Cost Each Year of Vanishing Rainforest," *The Independent*, October 3, 2010.

109. Ciara Raudsepp-Hearne et al., "Untangling the Environmentalist's Paradox: Why is Human Well-being Increasing as Ecosystem Services Degrade?" *Bioscience* 60, 8 (September 2010), pp. 576–589.

110. Jim, "Desdemona at 2: The Environmentalist's Paradox," Desdemona Despair, posted December 2, 2010.

111. European Environment Agency, *European Environment State and Outlook Report 2010* (Denmark: EEA, 2010).

112. Mark Kinver, "Growing Demand for Resources 'Threatens EU Economy'," *BBC News*, posted November 30, 2010.

113. Nafeez Ahmed, "Study: Global Warming, Energy Shortages, Food Scarcity and Recession Could Cause Industrial 'Failure' in 10 Years," *OpEdNews.com*, posted October 5, 2010.

114. Ahmed, "Study: Global Warming, Energy Shortages, Food Scarcity and Recession."

Chapter Four

1. See H. C. Barnett and C. Morse, *Scarcity and Growth* (Baltimore: Johns Hopkins Press, 1963).

2. For an excellent discussion of whether the shift from cheap conventional oil to expensive non-conventional liquid fuels constitutes "substitution" in the economic sense, see David Murphy, "Does Peak Oil Even Matter?," *The Oil Drum*, December 17, 2010, theoildrum.com/node/7246#more.

3. Environmental Protection Agency, *Biofuels and the Environment: The First Triennial Report to Congress, Draft Report* EPA/600/R-10/183A, January 28, 2011.

4. David Pimentel and Tad W. Patzek, "Ethanol Production Using Corn, Switchgrass and Wood; Biodiesel Production Using Soybean and Sunflower," *Natural Resources Research* 14, no.1 (March 2005).

5. Hosein Shapouri, James Duffield, and Michael Wang, "The Energy Balance of Corn Ethanol: An Update," *US Department of Agriculture*, Office of Energy Policy and New Uses, Report No. 813, 2002.

6. David Pimental and Tad Patzek, "Ethanol Production Using Corn, Switchgrass and Wood: Biodiesel Production Using Soybean and Sunflower," *Natural Resources Research* 14, no.1 (March 2005); Adam Liska et al., "Improvements in Life Cycle Energy Efficiency and Greenhouse Gas Emissions of Corn Ethanol," *J. Industrial Ecology*, 13, no.1 (2009); Jason Hill et al., "Environmental, Economic, and Energetic Costs and Benefits of Biodiesel and Ethanol Biofuels," *Proceedings of the National Academy of Sciences* 103 (July 25, 2006). A recent paper by David Murphy on ethanol (Murphy et al., "New Perspectives on the Energy Return on (Energy) Investment (EROI) of Corn Ethanol," *Environment, Development and Sustainability* 13:179–202, 2011) calculates the 95 percent confidence interval surrounding the various estimates of EROEI of corn

ethanol that are found in the literature. The results are that the EROEI = 1.01 ± 0.2. Therefore we cannot say, with any certainty, that the EROEI of corn ethanol is above 1.

7. Charles A. S. Hall and the "EROI Study Team," "Provisional Results from EROI Assessments," *The Oil Drum*, posted April 8, 2008, theoildrum.com/node/3810.

8. "New Forecast Warns Oil Will Run Dry Before Substitutes Roll Out," *UC Davis News Service*, posted November 9, 2010; Helen Knight, "Green Machine: Markets Hint at 100-Year Energy Gap," *New Scientist online*, posted November 11, 2010.

9. Kurt Zenz House, "The Limits of Energy Storage Technology," *Bulletin of the Atomic Scientists*, January 20, 2009.

10. Aeldric, "The Networking of Resource Production: Do the Networks Give Us Warnings When They Are About to Fail?" *The Oil Drum*, posted September 22, 2010, anz.theoildrum.com/node/6974#more.

11. See, for example, "Iron Ores," *Highbeam Business*, 2011. business.highbeam.com/industry-reports/mining/iron-ores; and Chanyaporn Chanjaroen, "Escondida Copper Production Will Decline Up to 10% on Ore Grades, BHP Says," *Bloomberg*, posted August 25, 2010.

12. Ugo Bardi, "The Universal Mining Machine," *The Oil Drum*, posted January 24, 2008, europe.theoildrum.com/node/3451.

13. The difficulties of rapid adaptation to fuel supply shortfalls are explored in more depth in the Hirsch Report. Robert L. Hirsch, Roger Bezdek, and Robert Wendling, *Peaking of World Oil Production: Impacts, Mitigation, & Risk Management*, Report for the US Department of Energy, February 2005, netl.doe.gov/publications/other_toc.html.

14. Thanks to Mats Larsson, personal communication.

15. See for example Benjamin S. Cheng, "An Investigation of Cointegration and Causality Between Energy Consumption and Economic Growth," *Journal of Energy and Development*. 21, no.1 (Autumn 1995). However, see also Jaruwan Chontanawat, Lester C. Hunt, and Richard Pierse, "Does Energy Consumption Cause Economic Growth?: Evidence From a Systematic Study of Over 100 Countries," *Journal of Policy Modeling* 30, no.2 (2008).

16. US Energy Information Administration, "Annual Energy Outlook 2010 with Projections to 2035," eia.doe.gov/oiaf/aeo/otheranalysis/aeo_2010analysispapers/intensity_trends.html.

17. In recent years, the Chinese government has set a goal of increasing the energy efficiency of the nation's economy (producing more GDP growth per unit of energy); it has accomplished this at least partially through draconian power cuts to cities and factories. Leslie Hook, "China's Energy Drive: Back On Track," *Beyondbrics*, posted November 25, 2010, blogs.ft.com/beyond-brics/2010/11/25/chinas-energy-efficiency-drive-back-on-track/.

18. See Cutler J. Cleveland, "Energy Quality," *The Encyclopedia of Earth*, eoearth.org/article/Energy_quality#gen11.

19. David Stern, "Energy Mix and Energy Intensity," *Stochastic Trend*, posted April 17, 2010, stochastictrend.blogspot.com/2010/04/energy-mix-and-energy-intensity.html.

20. Cutler J. Cleveland, "Energy Quality, Net Energy and the Coming Energy Transition," in *Frontiers in Ecological Economic Theory and Application*, Jon D. Erickson and John M. Gowdy, eds. (Cheltenham, UK: Edward Elgar, 2007), pp. 268–284.

21. Cleveland, "Energy Quality, Net Energy, and the Coming Energy Transition," 7;

Kenneth S. Deffeyes, chapter 3 in *Beyond Oil: The View From Hubbert's Peak* (New York: Hill and Wang, 2005); David I. Stern, "Energy and Economic Growth in the USA: A Multivariate Approach," Energy Economics 15, no.2 (1993), pp. 137–150.

22. Cleveland, "Energy Quality," *The Encyclopedia of Earth* (online).

23. In 2009, the US imported nearly $300 billion in consumer goods from China. For information on the trade balance between the two nations see US Census Bureau, Foreign Trade Statistics, census.gov/foreign-trade/balance/.

24. Cleveland, "Energy Quality, Net Energy, and the Coming Energy Transition."

25. David Murphy and Charles A. S. Hall, "EROI, Insidious Feedbacks, and the End of Economic Growth," pre-publication, 2010.

26. Ernst von Weizsacker, Amory Lovins, and L. Hunter Lovins, *Factor Four: Doubling Wealth, Halving Resource Use — The New Report to Rome* (Sydney, AU: Allen & Unwin, 1998).

27. Paul Hawken, Amory Lovins, and L. Hunter Lovins, *Natural Capitalism: The Next Industrial Revolution* (London: Earthscan, 2000).

28. Apsmith, "What's Wrong With Amory Lovins?," *SciScoop Science*, posted September 17, 2005; Robert Bryce, "Green Energy Advocate Amory Lovins: Guru or Fakir?" *Energy Tribune*, posted November 12, 2007; Robert Bryce, "An Interview With Vaclav Smil," *robertbryce.com*, posted July 2007.

29. This is known as the Jevons Paradox or the Khazzoom-Brookes postulate. In one sense, the Jevons Paradox has fading relevance in a world of declining energy resource availability: in the future, efforts to increase efficiency will probably not lead to declining resource prices, merely to prices that are not rising as fast as they would without such efforts. However, another, somewhat analogous trend will come into play: In order for society to save large amounts of energy through efficiency, very large investments in new and more energy efficient infrastructure are required (railroads, electric cars, etc.). This build-up of new energy efficient infrastructure requires energy for the construction and production, which will increase the need for energy. William Stanley Jevons, *The Coal Question* (London: Macmillan, 1866); Harry D. Saunders, "The Khazzoom-Brookes Postulate and Neoclassical Growth," *The Energy Journal* (October 1, 1992).

30. Wikipedia, "Luminous Efficacy," accessed January, 2011; Wikipedia, "Light-Emitting Diode: Efficiency and Operational Parameters," accessed January 2011.

31. Alan S. Drake, "Electrified and Improved Railroads," in *An American Citizen's Guide to an Oil-Free Economy: A How-To Guide For Ending Oil Dependency*, posted on aspousa .org, October 25, 2010.

32. Richard Heinberg and Michael Bomford, *The Food and Farming Transition*, Post Carbon Institute, 2009, available online at postcarbon.org/food.

33. Mats Larsson, *The Limits of Business Development and Economic Growth: Why Business Will Need to Invest Less in the Future* (New York: Palgrave Macmillan, 2004).

34. Larsson, *The Limits of Business*, p. 121.

35. Larsson, *The Limits of Business*, p. 5.

36. Larsson, *The Limits of Business*, p. 2.

37. Larsson, *The Limits of Business*, p. 133.

38. However, Larsson more than made up for these omissions in his later book, *Global Energy Transformation: Four Necessary Steps to Make Clean Energy the Next Success Story* (New York: Palgrave Macmillan, 2009).

39. Mats Larsson, personal communication.

40. For a discussion of limits to Moore's law see Brooke Crothers, "Moore's Law Limit Hit by 2014?" *cnet news*, posted June 16, 2009.

41. "Current and projected rates of innovation might not be sufficient to improve or even maintain living standards in the face of still rapidly growing population, global warming, and other challenges of the 21st century," according to Canadian economist James Brander. Barrie McKenna, "Has Innovation Hit a Brick Wall?" *The Globe and Mail*, December 26, 2010.

42. Martin Lowson, "A New Approach to Sustainable Transport Systems," presented at the 13th World Clean Air and Environmental Congress, London, August 22–27, 2004; The conversion is: 0.55 MJ = 521.6 Btu; 1.609 km = 1 mi; therefore, 521.6 × 1.609 = 839; US Department of Energy, "Transportation Energy Data Book," 2007.

43. Rick Jervis, "Pipes, Pumps Trouble Big Easy," *USA Today*, posted December 16, 2010.

44. US Government Accountability Office, *Global Positioning System: Significant Challenges in Sustaining and Upgrading Widely Used Capabilities*, Report No. GAO-09-670T, May 7, 2009.

45. Vernon W. Ruttan, *Is War Necessary for Economic Growth?: Military Procurement and Technology Development* (New York: Oxford University Press, 2006).

46. In *The Party's Over*, I discussed these strategies by which humanity has enlarged its carrying capacity, summarizing a longer seminal discussion in William Catton's brilliant 1980 book *Overshoot*. Richard Heinberg, *The Party's Over* (Gabriola Island BC: New Society Publishers, 2003), pp. 14–33; William Catton, *Overshoot: The Ecological Basis of Revolutionary Change* (University of Illinois, 1980).

47. For an unrealistically optimistic view of what division of labor has achieved, see John Kay, "Why You Can Have an Economy of People Who Don't Sweat," *Financial Times*, Oct. 20, 2010.

48. Adam Smith, *The Wealth of Nations* (New York: Oxford University Press, World's Classics Edition, 2008).

49. Economists frame the advantages of increased trade in terms of a concept called "comparative advantage." Technically, comparative advantage means that each country will specialize in what it is good at, even if another country does everything better, and the result is that all countries benefit from trade. However, this assumes that capital can only be invested domestically. If instead capital can be invested anywhere, which is actually the case, then it will go to whichever country has an absolute advantage. The net result is even greater output than a world based on comparative advantage, but one in which not all countries benefit. Even in the comparative advantage model, economists recognize that some sectors in a country can lose out while others gain. In world of absolute advantage, the sectors that gain need not be in the same country as the sectors that lose. Herman Daly has made this point, but apparently Ricardo also recognized that comparative advantage depends on capital being invested domestically, which was the case in his day.

50. Frances Berdan, "Trade and Markets in Precapitalist States," in *Economic Anthropology*, Stuart Plattner, ed. (Stanford, CA: Stanford University Press, 1989).

51. Jeff Rubin, *Why Your World Is About To Get a Whole Lot Smaller: Oil and the End of Globalization* (New York: Random House, 2009).

52. Jeff Rubin, "Oil and the End of Globalization," transcript and video of a presentation at 2010 ASPO-USA, *The Oil Drum*, posted November 8, 2010, theoildrum.com/node/7095#more.

53. Julian Lincoln Simon, *The Ultimate Resource* (Princeton, NJ: Princeton University Press, 1996).

54. Herman Daly, "Ultimate Confusion The Economics of Julian Simon," *Social Contract Journal* 13, no.3 (Spring, 2003).

55. Bjorn Lomborg, *The Skeptical Environmentalist* (Cambridge: Cambridge University Press, 2001); Bjorn Lomborg, "Cool It:" *The Skeptical Environmentalist's Guide to Global Warming* (New York: First Vintage Books, 2007); *Cool It*, movie by Ondi Timoner and Terry Botwick, 1019 Films, 2011.

56. The literature on the collapse of civilizations includes materials ranging widely in terms of quality of analysis. The best-selling recent book on the subject is Jared Diamond, *Collapse: How Societies Choose to Fail or Succeed* (New York: Viking, 2005).

57. Joseph A. Tainter, *The Collapse of Complex Societies*, New Studies in Archaeology, (Cambridge UK: Cambridge University Press, 1988).

Chapter Five

1. "Super Efficient Crew Builds 15-Story Chinese Hotel in Just SIX DAYS," *The Huffington Post*, posted November 12, 2009.

2. Richard Heinberg and David Fridley, "The End of Cheap Coal," *Nature* 468 (November 18, 2010), pp. 367–369.

3. Zaipu Tao and Mingyu Li, "What Is the Limit of Chinese Coal Supplies — A STELLA Model of Hubbert Peak," *Energy Policy* 35, no.6 (June, 2007).

4. Werner Zittel and Jorg Schindler, *Coal: Resources and Future Production*, EWG-Paper 1/07 (Ottobrunn, Germany: Energy Watch Group, March 28, 2007).

5. Tadeusz W. Patzek and Gregory D. Croft, "A Global Coal Production Forecast with Multi-Hubbert Cycle Analysis," *Energy* 35 (2010).

6. Don Miller, "Uranium Prices Surge on China's $511 Billion Investment in Nukes," Money Morning.com, posted December 7, 2010.

7. Elisabeth Rosenthal, "Nations That Debate Coal Use Export It to Feed China's Need," *The New York Times*, November 21, 2010.

8. This section draws heavily on the article "The Japan Syndrome" by Ethan Devine. However, see also the work of Chalmers Johnson on "the developmental state." Ethan Devine, "The Japan Syndrome," *Foreign Policy*, September 30, 2010; Chalmers Johnson, *MITI and the Japanese Miracle: The Growth of Industrial Policy, 1925–1975* (Stanford, CA: Stanford University Press, 1982).

9. Devine, "The Japan Syndrome," *Foreign Policy*.

10. From a social standpoint, Japan handled its economic transition away from high growth rates well: throughout the 1990s, the Japanese unemployment rate was about three percent, only half the US rate at the time. The Japanese also maintained universal healthcare, and had low wealth inequality, low rates of infant mortality, crime and incarceration, as well as the highest life expectancy. Steven Hill, "Reconsidering Japan and Reconsidering Paul Krugman," *Truth-out.org*, posted December 12, 2010.

11. Devine, "The Japan Syndrome," *Foreign Policy*.

12. Ted C. Fishman, "As Populations Age, a Chance for Younger Nations," *The New York Times*, October 14, 2010.

13. For another overview of this subject, see Dalibor Rohac, "Shaky Foundations to China's Growth," Asia Times online, posted October 21, 2010.

14. Chandni Rathod and Gus Lubin, "And Now Presenting: Amazing Satellite Images of the Ghost Cities of China," *Business Insider*, posted December 14, 2010.

15. Wieland Wagner, "China's Real Estate Bubble Threatens to Burst," *Spiegel Online*, posted August 3, 2010.

16. To read more, see Bill Powell, "Inside China's Runaway Building Boom," *Time.com*, posted April 5, 2010.

17. Warren Karlenzig, "China's New National Plan: Green By Necessity," *Common Current.com*, posted November 29, 2010.

18. The history of currencies is recounted in Niall Ferguson, *The Ascent of Money: A Financial History of the World* (New York: Penguin Press, 2008).

19. See Wikipedia.org, "Genoa Conference (1922)."

20. See Wikipedia.org, "History of the United States Dollar" and "Nixon Shock."

21. David E. Spiro, *The Hidden Hand of American Hegemony: Petrodollar Recycling and International Markets* (Ithaca, NY: Cornell University Press, 1999).

22. See Marvin Friend, "A Short History of US Monetary Policy," The Powell Center, powellcenter.org/econEssay/history.html.

23. Bill Black, "The EU's New Bailout Plan Will Exacerbate Political Crises," *Business Insider*, posted December 13, 2010.

24. See Wikipedia.org, "Foreign Exchange Market."

25. To read more, see Edward Harrison, "Currencies Pegged to the Dollar Under Pressure to Drop Peg," *Credit Writedowns.com*, posted October 13, 2009; Michael Hudson, "Why the US Has Launched a New Financial World War—And How the Rest of the World Will Fight Back," *alternet.org*, posted October 12, 2010.

26. Michael Sauga and Peter Müller, "The US Has Lived on Borrowed Money For Too Long," an interview with German Finance Minister Schäuble, *Spiegel online*, posted November 8, 2010.

27. See for example, "Finance Has Become a New Form of Warfare," an interview with Michael Hudson, *Information Clearing House*, posted November 5, 2010; David De-Graw, "The Road to World War III—The Global Banking Cartel Has One Card Left to Play," *Amped Status*, posted September 23, 2010; Ambrose Evans-Pritchard, "QE2 Risks Currency Wars and the End of Dollar Hegemony," *The Telegraph*, November 1, 2010.

28. Ambrose Evans-Pritchard, "The Eurozone is in Bad Need of an Undertaker," *The Telegraph*, December 12, 2010.

29. Michael Klare, *Rising Powers, Shrinking Planet: The New Geopolitics of Energy* (New York: Henry Holt, 2008), chapter 6.

30. Vicken Cheterian, "The Arab Crisis: Food, Energy, Water, Justice," *Energy Bulletin*, posted January 26, 2011.

31. Brice Pedroletti, "China Seeks to Mine Deep Sea Riches," *The Guardian*, December 7, 2010.

32. For an excellent general overview of the population issue, see Bill Ryerson, "Population: The Multiplier of Everything Else," in *The Post Carbon Reader* (Healdsburg, CA: Watershed Media, 2010); for an interactive Internet guide to population sizes and growth rates for all regions and countries, see "Explore the US Census Bureau's International Data Base," *mazamascience.com*, mazamascience.com/Population/IDB/.

33. "Pakistan: Population growth rate adds to problems," irinnews.org/Report.aspx?ReportID=91656 (accessed March 16, 2011).

34. Spengler, "Longevity Gives Life to Tea Party," *Asia Times online*, posted December 7, 2010.

35. Population Media Center, populationmedia.org.

36. Sharon Astyk, *Depletion and Abundance: Life on the New Home Front* (Gabriola Island, BC: New Society Publishers, 2008); Sharon Astyk, "Peak Oil Is a Women's Issue," *Relocalize.net*, posted November 25, 2006.

37. "Mother: Caring Our Way Out of the Population Dilemma," by Christophe Fauchere and Joyce Johnson, Tiroir A Films, 2011, motherthefilm.com.

38. See for example Robin Broad and John Cavanagh, *Development Redefined: How the Market Met Its Match* (Boulder, CO: Paradigm Publishers, August 2008); Edward Goldsmith et al., *The Future of Progress: Reflections on Environment and Development* (Totnes, Devon: Green Books, 1995).

39. Helena Norberg-Hodge, *Ancient Futures: Learning From Ladakh* (San Francisco: Sierra Club Books, 1991); John Perkins, *Confessions of an Economic Hit Man* (New York: Penguin Group, 2006).

40. United Nations Human Development Report 2010, *The Real Wealth of Nations: Pathways to Human Development* (New York: Palgrave Macmillan, November, 2010).

41. See the report by Reset for the House of Commons All Party Parliamentary Committee on Peak Oil, *The Impact of Peak Oil on Economic Development* (Edinburgh: Reset, 2008).

42. Amartya Sen, *Development as Freedom* (New York: Anchor Books, 1999).

43. Amartya Sen, "Development as Capability Expansion," in *Readings in Human Development*, Sakiko Fukuda-Parr and A. K. Shiva Kumar, eds. (New York: Oxford University Press), pp. 3–16.

44. See page 22, "Quantitative Relationships Among Energy Use, GDP, and Other Socionomic Indicators," in James H. Brown et al., "Energetic Limits to Economic Growth," *Bioscience* 61, no.1 (January 2011), pp. 19–26.

45. Wikipedia, "Overdevelopment," en.wikipedia.org/wiki/Overdevelopment.

46. Manfred A. Max-Neef, Antonio Elizalde, and Martin Hopenhayn, "Development and Human Needs," chapter in *Human Scale Development: Conception, Application, and Further Reflections* (New York: Apex, 1991), p. 18.

47. Manfred Max-Neef, Antonio Elizalde, and Martin Hopenhayn, (in Spanish) "Desarrollo a Escala Humana — Una Opción Para el Futuro," *Development Dialogue*, número especial (CEPAUR y Fundación Dag Hammarskjold, 1986), p. 12; Manfred Max-Neef et al., "Human Scale Development: An Option for the Future," *Development Dialogue* 1 (1989), pp. 7–80.

48. Wikipedia, "International Inequality," en.wikipedia.org/wiki/International_inequality.

49. Jim, "Desdemona at 2: The Environmentalist's Paradox," *Desdemona's Despair*, posted December 2, 2010.

50. James B. Davies et al., *The World Distribution of Household Wealth*, United Nations University's World Institute for Development Economics Research (New York: UNU-WIDER, February 2008).

51. Naomi Klein, *The Shock Doctrine: The Rise of Disaster Capitalism* (New York: Picador, 2007).

52. (January/February 2011); Robert B. Reich, *Aftershock: The Next Economy and America's Future* (New York: Alfred A. Knopf, 2010).

Chapter 6

1. Michael Cooper and Mary Williams Walsh, "Mounting Debts by States Stoke Fears of Crisis," *The New York Times*, December 4, 2010.
2. The connection between falling standards of living and political unrest has long been a subject of study. See J. C. Davies, "The J Curve of Rising and Declining Satisfactions As a Cause of Some Great Revolutions and a Contained Rebellion," in *The History of Violence in America: Historical and Comparative Perspectives*, H. D. Graham and T. R. Gurr, eds. (New York: Praeger, 1969), pp. 690–730.
3. Dana Priest and William M. Arkin, "Monitoring America," *The Washington Post*, December 20, 2010.
4. Rick Munroe, "Review: Putting the Bundeswehr Report in Context," *Energy Bulletin*, posted September 28, 2010.
5. Ferguson is discussing government debt only, but I have reworded his points somewhat and extrapolated them to refer to a more general debt crisis. Niall Ferguson, "Fiscal Crises and Imperial Collapses: Historical Perspectives on Current Predicaments," presentation posted online, *Business Insider*, May 19, 2010, businessinsider.com/niall-ferguson-sovereign-debt-2010-5#-26.
6. Richard Douthwaite and Gillian Fallon, eds., *Fleeing Vesuvius: Overcoming the Risks of Economic and Environmental Collapse* (Gabriola Island, BC: New Society Publishers, forthcoming), p. 76.
7. Ellen Brown, "QE2: It's the Federal Debt, Stupid!" *Truthout*, posted November 12, 2010.
8. Ellen Brown, "Austerity Fails in Euroland: Time for Some 'Deficit Easing,'" *Truthout*, posted December 24, 2010.
9. A similar strategy is being pursued by JAK bank in Sweden, which issues no-interest loans to customers.
10. Ellen Brown, "Cash-Starved States Need to Play the Banking Game: North Dakota Shows How," *The Web of Debt*, posted March 2, 2009.
11. The government could loan money into existence interest-free for projects deemed to be in the public interest that would generate enough revenue to repay the debt. It could buy municipal bonds at zero percent interest, for example, to help the current municipal fiscal crunch. Interest is the culprit more than the debt-based creation of money.
12. See, for example, Stephen Zarlenga's *The Lost Science of Money* (New York: American Monetary Institute, 2002) and Margrit and Declan Kennedy's *Interest and Inflation Free Money* (Lansing MI: Seva International, 1995).
13. Herman E. Daly and Joshua Farley, *Ecological Economics* (Washington DC: Island Press, 2004), p. 258.
14. Unfortunately, however, banks loan when the economy is booming and collect payments when it is shrinking, the opposite of what is required.
15. Transaction Net, "LETS — Local Exchange Trading Systems," transaction.net/money/lets/.
16. Dierdre Kent, *Healthy Money, Healthy Planet* (Nelson, NZ: Craig Potton, 2005); Richard Douthwaite, *The Ecology of Money* (Totnes, Devon: Green Books, 2000); Bernard Lietaer, *The Future of Money: Creating New Wealth, Work and a Wiser World* (London: Random House, 2001); and Thomas Greco, Jr., *Money: Understanding and Creating Alternatives to Legal Tender* (White River Junction, VT: Chelsea Green, 2001).

17. BerkShares, Inc., "Local Currency for the Berkshire Region," berkshares.org.

18. Other monetary theorists, including Bernard Lietaer, Richard Douthwaite, and Ellen Brown have arrived at essentially the same conclusion. See Ellen Brown, "Vision: Time for New Theory of Money," *YES! Magazine*, posted December 21, 2010; Lietaer, *The Future of Money*; Richard Douthwaite, "The Supply of Money in an Energy-Scarce World," in *Fleeing Vesuvius*, pp. 58–83.

19. Bernard Lietaer, "Why This Crisis? And What to Do About It?" lecture, TEDx Berlin, November 15, 2010, online at lietaer.com/; Bernard Lietaer, Robert Ulanowicz, and Sally Goerner, "Managing the Banking Crisis," *Lietaer.com*, posted February 4, 2010.

20. Greco, *The End of Money*, pp. 171–189.

21. Greco, *The End of Money*.

22. Upon reading this paragraph, ecological economist Josh Farley commented: "My personal view is that the state should be solely responsible for creating money. It would be a receipt/IOU for services rendered, repayable in the form of taxes. Taxes would be used to regulate the money supply rather than a means of financing government expenditure. This would allow money creation by federal, state and local governments. To get around the Constitution, state and local governments would issue interest free bonds accepted for taxes rather than legal tender. I've participated in enough complementary currencies and credit clearing systems to recognize that their reality often falls short of their theoretical beauty."

23. "Bartering, an age-old mode of commerce, has taken hold with a vengeance this year as the recession draws a broader spectrum of people trading everything from designer clothes to guitar lessons. The phenomenon is rooted partly in environmental concerns about crowded landfills and the energy consumed in manufacturing as well as a mainstream embrace of recycling. Social media like Facebook lend momentum to the swaps as people join forces to trade, share or negotiate better deals from retailers." Mireya Navarro, "In a Tight Holiday Season, Some Turn to Barter," *The New York Times*, December 22 2010.

24. Andrew Gavin Marshall, "The Financial New World Order: Towards a Global Currency and World Government," *Global Research.ca*, posted April 6, 2009.

25. See, for example, Daly and Farley, *Ecological Economics*.

26. When a non-renewable resource is essential and substitution is very difficult, depletion should proceed no faster than we can develop renewable substitutes. We should capture the unearned income from the extraction of such resources (also known as rent or windfall profits, e.g. Exxon's $46 billion in profits in 2008) and invest it in renewable substitutes.

27. Another way to say this: Waste emissions cannot exceed waste absorption capacity.

28. Lietaer, TEDx Berlin lecture, on lietaer.com. As economists define efficiency, it boils down to a maximization of monetary value, determined by willingness to pay. Converting corn to ethanol for Americans to drive their SUVs is more efficient than using it to feed malnourished Mexicans if Americans are willing to pay more. Perhaps we need to redefine efficiency as the satisfying the greatest number of human needs at the lowest ecological cost. Since GDP is a good proxy for ecological costs, it should go in the denominator of national welfare accounts.

29. Organizations studying and promoting ecological economics include: The International Society of Ecological Economics, The Center for the Advancement of the Steady

State Economy, The New Economics Institute, New Economics Foundation, The Capital Institute, and The Gund Institute.

30. Tim Jackson, *Prosperity Without Growth* (London: Earthscan, 2009).

31. D. Woodward and A. Simms, *Growth Isn't Possible* (London: New Economics Foundation, 2006), available online at neweconomics.org.

32. Herman Daly, "A Shift in the Burden of Proof," *The Daly News*, steadystate.org/a-shift-in-the-burden-of-proof/.

33. Herman Daly, "Uneconomic Growth in Theory and in Fact," the first annual Feasta lecture, at Trinity College Dublin, April 26, 1999.

34. Serge Latouche, *Farewell to Growth* (Malden, MA: Polity Press, 2009).

35. Frederick Soddy, *Wealth, Virtual Wealth and Debt* (London: George Allen & Unwin, 1926).

36. Henry David Thoreau, *Walden* (Boston, 1854); Scott Nearing and Helen Nearing, *Living the Good Life* (New York: Schocken Books, 1970).

37. Duane Elgin, *Voluntary Simplicity* (New York: HarperCollins, 1981); Joe Dominguez and Vicki Robin, *Your Money or Your Life* (New York: Viking Penguin, 1992).

38. Henry George, *Progress and Poverty* (New York: Doubleday, 1879); The Henry George Institute.

39. Thorstein Veblen, *The Theory of the Leisure Class: An Economic Study of Institutions* (New York: Macmillan, 1899).

40. Some other organizations and websites include the New Economics Institute, Capital Institute, New Economics Foundation, Third Millennium Economy, Real World Economics, paecon.net/PAEReview/, and ethicalmarkets.com.

41. Herman Daly, *Steady-State Economics* (Washington, DC: Island Press, 1991), p. 17.

42. Center for the Advancement of a Steady State Economy (CASSE), steadystate.org.

43. Brian Czech, *Shoveling Fuel for a Runaway Train* (Berkeley: University of California Press, 2000).

44. Peter Victor, *Managing Without Growth: Slower by Design, Not Disaster* (Cheltenham, UK: Edward Elgar, 2008); also Rob Dietz, "The 'Good New Days' in a Non-Growing Economy," *Energy Bulletin*, posted October 25, 2010.

45. Susan Arterian Chang, "Moving Towards a Steady-State Economy," *The Finance Professionals' Post*, a publication of the New York Society of Security Analysts, posted April 12, 2010.

46. "Eco-Municipalities: Sweden and the United States," *knowledgetemplates.com*, knowledgetemplates.com/sja/ecomunic.htm.

47. Daly and Farley, *Ecological Economics*, pp. 178–179.

48. Daly and Farley, *Ecological Economics*, p. 180.

49. Douthwaite and Fallon, *Fleeing Vesuvius*, p. 142 and following.

50. Douthwaite and Fallon, *Fleeing Vesuvius*, pp. 89–111; Daniel Fireside, "Community Land Trust Keeps Prices Affordable — For Now and Forever," I, posted July 29, 2008; "Land Trust Success," *Land Trust Alliance*, landtrustalliance.org. Brazil already does this, and has also imposed an additional tax on foreign investments in their stock market.

51. Hazel Henderson, hazelhenderson.com.

52. Hazel Henderson, "Real Economies and the Illusions of Abstraction," *Energy Bulletin*, posted December 8, 2010; CDFI: Coalition of Community Development Financial Organizations, cdfi.org.

53. Joel Bakan, "The Corporation," documentary movie, Big Picture Media Corporation, 2003.
54. Green America, greenamerica.org.
55. International Co-operative Alliance, "Statement on the Co-operative Identity," ica .coop/coop/principles.html.
56. For expanded discussion of these points, see discussion of GNP in Arne Naess, *Ecology, Community, and Lifestyle: Outline of an Ecosophy* (Cambridge, UK: Cambridge University Press, 1989).
57. William D. Nordhaus and James Tobin, *Is Growth Obsolete?*, Cowles Foundation Discussion Paper 319 (New Haven: Yale University, 1971).
58. Unfortunately, the organization Redefining Progress seems to have become a casualty of the economic crisis.
59. Harvard Medical School Office of Public Affairs, "Happiness is a Collective — Not Just Individual — Phenomenon," news alert, web.med.harvard.edu/sites/RELEASES/html /christakis_happiness.html.
60. Jamie Smith Hopkins, "Putting a Dollar Figure on Progress," *The Baltimore Sun*, September 11, 2010.
61. Joseph Stiglitz, Commission on the Measurement of Economic Performance and Social Progress, *Report on the CMEPSP* (September, 2009), p. 7.
62. The phrase "Green National Product" is from Clifford Cobb and John Cobb, *The Green National Product: A Proposed Index of Sustainable Economic Welfare* (Minnesota: University Press of America, 1994), pp. 280–281.
63. Derek Bok, *The Politics of Happiness: What Government Can Learn From the New Research on Well-Being* (Princeton: Princeton University Press, 2010).
64. "David Cameron Aims to Make Happiness the New GDP," *The Guardian*, November 14, 2010.
65. "Seattle Area Happiness Initiative," *Sustainable Seattle.org*, sustainableseattle.org.
66. "ABAC Poll: Thai People Happiness Index Rose to 8 Out of 10 Points," eThailand.com, posted December 6, 2010.
67. "Coronation Address of His Majesty King Khesar, the 5th Druk Gyalpo of Bhutan," Gross National Happiness.com, November 7, 2008.
68. Cliff Kuang, "Infographic of the Day: Happiness Comes at a Price," fastcodesign.com, posted December 8, 2010; happyplanetindex.org.
69. Helena Norberg-Hodge, Steven Gorelick, and John Page, "The Economics of Happiness," a documentary movie, International Society for Ecology & Culture, 2011.
70. Lester Brown, *Plan B 4.0: Mobilizing to Save Civilization* (New York: W. W. Norton, 2009).
71. Lester Brown, *World on the Edge: How to Prevent Environmental and Economic Collapse* (New York: W. W. Norton, 2011).
72. The book also discusses Tradable Energy Quotas, a consumer- and market-driven cap-and-trade fossil fuels quota rationing scheme now being evaluated by Parliament in the UK. Richard Heinberg, *The Oil Depletion Protocol: A Plan to Avert Oil Wars, Terrorism and Economic Collapse* (Gabriola Island, BC: New Society, 2006). See also David Fleming and Shaun Chamberlin, "TEQs Tradable Energy Quotas: A Policy Framework for Peak Oil and Climate Change," *Energy Bulletin*, posted January 18, 2011; TEQs: Media Coverage and Launch, *teqs.net*, teqs.net/report/media-coverage-and -launch-event/.

73. Albert Bates, *The Biochar Solution* (Gabriola Island, BC: New Society, 2010).

74. Alan S. Drake, "Chapter 1: Electrified and Improved Railroads," from *An American Citizen's Guide to an Oil-Free Economy*, Energy Bulletin, posted October 27, 2010.

75. William N. Ryerson, "Population: The Multiplier of Everything Else," in *The Post Carbon Reader*, Richard Heinberg and Daniel Lerch, eds. (Healdsburg, CA: Watershed Media, 2010), p. 153.

76. Roy Morrison, *Ecological Democracy* (Cambridge, MA:South End Press, 1995).

77. Joe Roman, Paul R. Ehrlich, Robert M. Pringle, and John C. Avise, "Facing Extinction: Nine Steps to Save Biodiversity," *Solutions* 1, no.1 (February 24, 2009), pp. 50–61.

78. Richard Heinberg and Michael Bomford, "The Food and Farming Transition," available online at Post Carbon.org, 2009.

79. The cultural, psychological, and institutional barriers to solving our 21st-century survival challenges are being addressed by the Millennium Assessment of Human Behavior. Stanford University, "Millenium Assessment of Human Behavior," mahb .stanford.edu.

80. This is a slight rephrasing of Winston Churchill's famous remark that "[t]he United States invariably does the right thing, after having exhausted every other alternative."

81. Robert Ornstein and Paul Ehrlich, *New World New Mind: Moving Toward Conscious Evolution* (Cambridge, MA: Malor Books, 2000).

82. Peter C. Whybrow, *American Mania: When More is Not Enough* (New York: W. W. Norton, 2005).

83. Nate Hagens, "The Psychological and Evolutionary Roots of Resource Overconsumption Revisited," *The Oil Drum*, posted June 25, 2009, theoildrum.com/node/5519.

84. Kurt Cobb, "The Extremely Leisurely Pace of American Democracy and the Urgency of Our Predicament," *Resource Insights*, posted January 9, 2011.

85. There are long and deep traditions valorizing voluntary poverty in Christian, Muslim, and Buddhist societies and literature. Many pre-agricultural societies were organized around a "Big Man" who gained prestige by giving away all his material possessions. See Marvin Harris and Orna Johnson, *Cultural Anthropology*, 7th ed. (Pearson, 2006), pp. 172–175.

86. Whybrow, *American Mania*.

Chapter 7

1. Rebecca Solnit, *A Paradise Built in Hell: The Extraordinary Communities That Arise in Disaster* (New York: Viking Books, 2009).

2. Lewis Aptekar, *Environmental Disasters in Global Perspective* (New York: G. K. Hall, 1994).

3. See James Howard Kunstler, *The Geography of Nowhere: The Rise and Decline of America's Man-Made Landscape* (New York: Free Press, 1994); and Gregory Greene, "The End of Suburbia," video documentary, The Electric Wallpaper Co., 2004.

4. Wikipedia, "Transition Towns," en.wikipedia.org/wiki/Transition_Towns.

5. Rob Hopkins, *Transition Handbook* (White River Junction, VT: Chelsea Green, 2008).

6. Hopkins, *Transition Handbook*, p. 15.

7. Formerly archived at: transitiontowns.org/TransitionNetwork/CheerfulDisclaimer

8. *Transition Town Totnes*, transitiontowntotnes.org.

9. Rob Hopkins, "Ingredients of Transition: Strategic Local Infrastructure," *Energy Bulletin*, posted December 15, 2010.

10. *Common Security Clubs*, commonsecurityclub.org.
11. "Discover Tools for Facilitators," and "Common Security Club Forum," *Common Security Clubs*, commonsecurityclub.org.
12. Sarah Byrnes, "Learning to Live on Less," *YES! Magazine*, posted December 23, 2010.
13. *Community Action Partnership*, communityactionpartnership.com.
14. I suggest the term "laboratory" because the sorts of efforts and enterprises that will best serve communities under rapidly evolving economic circumstances may not be apparent or even knowable at the outset — we will have to experiment. However, it is by no means essential or even important that the entities envisioned adopt this suggested title. Some communities may prefer slightly different names for political reasons: a Local Enterprise Laboratory, for example, might fare better in red states.
15. Colleen Kimmet, "Better Than A Food Bank," *Energy Bulletin*, posted November 5, 2010; *Mission Mountain Food Enterprise Center*, mmfec.org.
16. Marcin Jakubowski, "Global Village Construction Set," *Make*, Green Project Contest, makezine.com/tagyourgreen/detail.csp?id=95.
17. "The JAK Members Bank," *JAK Medlemsbank*, jak.aventus.nu/22.php.
18. "Sustainable Commercial Urban Farm Incubator (SCUFI) Program," Virtually Green .com.
19. John Michael Greer, *The Ecotechnic Future: Envisioning a Post-Peak World* (Gabriola Island, BC: New Society, 2009), p. 76.
20. *Dancing Rabbit Ecovillage*, dancingrabbit.org.
21. Diana Leafe Christian, *Finding Community: How to Join an Ecovillage or Intentional Community* (Gabriola Island, BC: New Society, 2007) and *Creating a Life Together: Practical Tools to Grow Ecovillages and Intentional Communities* (Gabriola Island, BC: New Society, 2003); *Communities: Life in Cooperative Culture Magazine*, communities. ic.org; *Global Ecovillage Network*, gen.ecovillage.org.
22. You can also access a trove of articles about them through *YES!* Magazine.
23. The phrase "great turning" has been used by activist and Buddhist teacher Joanna Macy, and was adopted by David Korten as the title of a recent book. The "turning" I am referring to is perhaps less politically and spiritually nuanced than what Macy and Korten describe.
24. Richard Wrangham, *Catching Fire: How Cooking Made Us Human* (New York: Basic Books, 2009).

Index

Page numbers in *italics* indicate figures.

energy alternatives: in China, 192–193; effect of oil prices, 16–17; EROEI, 116; food use, 133; forecasts, 171; as fossil fuel replacement, 115, 117–118, 148, 150, 157–162; resource needs, 161
Energy Descent Action Plan, 272
Energy Information Administration, 164
Energy Intensity Ratio (EIR), 123, 124
Energy Research Institute, 191
Energy Watch Group, 143, 192
The Entropy Law and the Economic Process (Georgescu-Roegen), 248–249
environment: asset price, 151–152; damage as externality, 41–42, 252; declines and human welfare, 149; human influence, 145; indirect services, 150–151; solutions to resource depletion, 259–266; threats to economic growth, 2, 152–153
Environment State and Outlook Report, 151
environmental disasters, 4, 146–148, 151–152
Environmental Disasters in Global Perspective (Aptekar), 269
environmental economists, 247
Environmental Research Letters (King), 123
environmentalist`s paradox, 149
EROEI (energy returned on energy invested), 112, 116, 118, 119, 158
Esteva, Gustavo, 217
ethanol, 134, 157–158, 171
Ethical Markets (Henderson), 253
Ethiopia, 129
European Central Bank, 45, 92, 239

European Commission, 91
European Environmental Agency, 151
European Union, 66, 204, 206, 209, 226
exports, 166, 193–196
externalities, 40–41, 252
ExxonMobil, 111, 156

F
Factor Four (Lovins), 168, 170
Factor 10, 170
family planning, 212–215
Fannie Mae, 58, 69, 84–85, 89, 96
Farley, Josh, 252
Federal government: debt, 73–75, 81–83; effect on deflation, 99–100; funding debt, 94; money creation, 239–240; response to financial collapse, 236–241; role in economy management, 45, 234–235; stimulus and bailout packages, 8, 9, 10, 20, 66, 79–80, 83–86, 87, 88, 89
Federal Housing Finance Agency, 84
Federal Reserve: changes in global currency, 245; deflation, 61, 99–100; financial bailouts, 89–90; interest rate changes, 56, 63; money creation, 240; new roles, 90–91, 95; quantitative easing, 75, 90, 94, 240; role of, 44
Ferguson, Niall, 237
fertility, 212–215
financial crisis (19th century), 44
financial crisis (2008), 8, 27, 63–64, 66, 68–69, 70–71
Financial Crisis Inquiry Commission, 68–69
financial institutions: impact of Gulf oil spill, 3; insolvency, 80–81; international actions, 92; invention of

banking, 31; potential collapse of, 233; regulations, 18, 44, 58, 60; state-owned banks, 240–241
financial services industry: bailout, 66, 83–85, 89–90; collapse in, 63, 64, 66, 68–69; corporate lending, 77; debt, 78–81; effect of, 46–51; effect of deregulation, 56; foreclosures, 95–96; fraud, 79–80; product risk, 60
financial-monetary system: alternatives, 241–245; crisis, 18–19; diversion of funds, 112; effect of mortgage changes, 62–63; effect of shadow banking system, 61–62; global currency, 245–246; government regulation, 45; impact of economic growth to, 6–7; potential collapse of, 233–236; recession recovery, 8, 10; response to collapse, 236–241; source of money, 80; sustainability of, 231–232; threats to economic growth, 2–3
Finland, 251
fire, harnessing of, 284
fisheries, 3, 135–137, 153
Fitch, 60
Fleeing Vesuvius (Douthwaite), 239
floods, 146, 147
"The Food and Farming Transition" (Heinberg and Bomford), 260
food riots, 19
food system, 129–138, 173, 272
Ford, Henry, 242
foreclosures, 65, 95–96
foreign investment, 61
Foreign Policy, 194
fossil fuels, 7, 11–12, 109, 150. See also specific fuels
Foxconn Technology Group, 197

About the Author

RICHARD HEINBERG is the author of nine award-winning books including *The Party's Over, Powerdown, Peak Everything*, and *Blackout*. He is a Senior Fellow of the Post Carbon Institute and is widely regarded as one of the world's foremost Peak Oil educators. He is a recipient of the M. King Hubbert Award for excellence in energy education, and since 2002 has given over 400 lectures on fossil fuel depletion to audiences around the world. He has been published in *Nature*, the world's premiere

scientific journal; he has been quoted in *Time* magazine; he has been featured in television and theatrical documentaries including Leonardo DiCaprio's "11th Hour"; and he has been interviewed on national radio and television in seven countries. To learn more about Richard and his work, visit TheEndOfGrowth.com

If you have enjoyed *The End of Growth,*
you might also enjoy other

BOOKS TO BUILD A NEW SOCIETY

Our books provide positive solutions for people who want to
make a difference. We specialize in:

Sustainable Living • Green Building • Peak Oil
Renewable Energy • Environment & Economy
Natural Building & Appropriate Technology
Progressive Leadership • Resistance and Community
Educational & Parenting Resources

New Society Publishers

ENVIRONMENTAL BENEFITS STATEMENT

New Society Publishers has chosen to produce this book on recycled paper made
with **100% post consumer waste,** processed chlorine free, and old growth free.
For every 5,000 books printed, New Society saves the following resources:[1]

35	Trees
3,126	Pounds of Solid Waste
3,439	Gallons of Water
4,486	Kilowatt Hours of Electricity
5,682	Pounds of Greenhouse Gases
24	Pounds of HAPs, VOCs, and AOX Combined
9	Cubic Yards of Landfill Space

---

[1]Environmental benefits are calculated based on research done by the Environmental Defense
Fund and other members of the Paper Task Force who study the environmental impacts of the
paper industry.

For a full list of NSP's titles, please call 1-800-567-6772 *or check out our website* at:

www.newsociety.com

NEW SOCIETY PUBLISHERS